# FRONTIERS IN THE STUDY OF CHAOTIC DYNAMICAL SYSTEMS WITH OPEN PROBLEMS

## WORLD SCIENTIFIC SERIES ON NONLINEAR SCIENCE

Editor: Leon O. Chua
University of California, Berkeley

---

**Series B. SPECIAL THEME ISSUES AND PROCEEDINGS**

Volume 1: Chua's Circuit: A Paradigm for Chaos
Edited by *R. N. Madan*

Volume 2: Complexity and Chaos
Edited by *N. B. Abraham, A. M. Albano, A. Passamante, P. E. Rapp, and R. Gilmore*

Volume 3: New Trends in Pattern Formation in Active Nonlinear Media
Edited by *V. Perez-Villar, V. Perez-Munuzuri, C. Perez Garcia, and V. I. Krinsky*

Volume 4: Chaos and Nonlinear Mechanics
Edited by *T. Kapitaniak and J. Brindley*

Volume 5: Fluid Physics — Lecture Notes of Summer Schools
Edited by *M. G. Velarde and C. I. Christov*

Volume 6: Dynamics of Nonlinear and Disordered Systems
Edited by *G. Martínez-Mekler and T. H. Seligman*

Volume 7: Chaos in Mesoscopic Systems
Edited by *H. A. Cerdeira and G. Casati*

Volume 8: Thirty Years After Sharkovskiĭ's Theorem: New Perspectives
Edited by *L. Alsedà, F. Balibrea, J. Llibre, and M. Misiurewicz*

Volume 9: Discretely-Coupled Dynamical Systems
Edited by *V. Pérez-Muñuzuri, V. Pérez-Villar, L. O. Chua, and M. Markus*

Volume 10: Nonlinear Dynamics & Chaos
Edited by *S. Kim, R. P. Behringer, H.-T. Moon, and Y. Kuramoto*

Volume 11: Chaos in Circuits and Systems
Edited by *G. Chen and T. Ueta*

Volume 12: Dynamics and Bifurcation of Patterns in Dissipative Systems
Edited by *G. Dangelmayr and I. Oprea*

Volume 13: Modeling and Computations in Dynamical Systems: In Commemoration of the 100th Anniversary of the Birth of John von Neumann
Edited by *E. J. Doedel, G. Domokos & I. G. Kevrekidis*

Volume 14: Dynamics and Control of Hybrid Mechanical Systems
Edited by *G. Leonov, H. Nijmeijer, A. Pogromsky & A. Fradkov*

Volume 15: From Physics to Control Through an Emergent View
Edited by *L. Fortuna, A. Fradkov & M. Frasca*

Volume 16: Frontiers in the Study of Chaotic Dynamical Systems with Open Problems
Edited by *Z. Elhadj & J. C. Sprott*

WORLD SCIENTIFIC SERIES ON NONLINEAR SCIENCE Series B Vol. 16

Series Editor: Leon O. Chua

# FRONTIERS IN THE STUDY OF CHAOTIC DYNAMICAL SYSTEMS WITH OPEN PROBLEMS

edited by

Elhadj Zeraoulia
University of Tébessa, Algeria

Julien Clinton Sprott
University of Wisconsin, Madison, USA

NEW JERSEY • LONDON • SINGAPORE • BEIJING • SHANGHAI • HONG KONG • TAIPEI • CHENNAI

*Published by*

World Scientific Publishing Co. Pte. Ltd.

5 Toh Tuck Link, Singapore 596224

*USA office:* 27 Warren Street, Suite 401-402, Hackensack, NJ 07601

*UK office:* 57 Shelton Street, Covent Garden, London WC2H 9HE

**British Library Cataloguing-in-Publication Data**
A catalogue record for this book is available from the British Library.

**World Scientific Series on Nonlinear Science, Series B — Vol. 16**
**FRONTIERS IN THE STUDY OF CHAOTIC DYNAMICAL SYSTEMS WITH OPEN PROBLEMS**



ISBN-13 978-981-4340-69-4
ISBN-10 981-4340-69-3

Printed in Singapore.

# Preface

*Anyone who has never made a mistake has never tried anything new! —Albert Einstein (1879–1955)*

*For every problem, there is one solution which is simple, neat and wrong! —H. L. Mencken (1880–1956)*

This collection of review articles is devoted to new developments in the study of chaotic dynamical systems with some open problems and challenges. The papers, written by many of the leading experts in the field, cover both the experimental and theoretical aspects of the subject. This volume presents a variety of fascinating topics of current interest and problems arising in the study of both discrete and continuous-time chaotic dynamical systems. Exciting new techniques stemming from the area of nonlinear dynamical systems theory are currently being developed to meet these challenges. This volume presents the state-of-the-art of the more advanced studies of chaotic dynamical systems.

*Frontiers in the study of chaotic dynamical systems with open problems* is devoted to setting an agenda for future research in this exciting and challenging field.

Zeraoulia Elhadj
Julien Clinton Sprott
October, 2010

## Contents

# Chapter 1

## Problems with Lorenz's Modeling and the Algorithm of Chaos Doctrine

Shoucheng OuYang[1] and Yi Lin[2,3]

[1]*Department of Scientific Research*
*Chengdu University of Information Technology*
*Chengdu, Sichuan 610041, PR China*
*E-mail: ouyangsc2000@yahoo.com.cn*

[2]*Department of Mathematics and Systems Science, College of Science*
*National University of Defense Technology*
*Changsha, Hunan, People's Republic of China*
[3]*Department of Mathematics, Slippery Rock University*
*Slippery Rock, PA 16057, USA*
*E-mail: Jeffrey.forrest@sru.edu*

In this paper, we first discuss the problems that exist in the modeling used in Lorenz's chaos theory by employing formal mathematical logic and the underlying physical meanings. Then we analyze in detail the problems found in computing Lorenz's model by employing the commonly employed schemes of computational mathematics. Our results indicate that the resultant chaos doctrine is not the chaos that appears in the physical events as Lorenz described; instead it involves the error values of the mathematical differences of quasi-equal quantities, producing the apparent chaos. Therefore, the problem of how to comprehend indeterminacy emerges.

**Keywords**: Mathematical model, nonlinearity, chaos, computational scheme, computational inaccuracy of quasi-equal quantities, indeterminacy.

### Contents

## 1.1. Introduction

Lorenz's chaos doctrine was initially published in the early 1960s [Lorenz, 1963] but did not receive much attention in the area of meteorology in the following twenty some years; especially, the front-line forecasters did not think much of it for the following reason:

**Problem 1.1.** *What is the fundamental reason behind the fact that the conventional theories of meteorology cannot practically resolve the problem of forecasting the forthcoming appearance of weather systems?*

Because of this reason, the front-line forecasters do not really follow these meteorological theories. Their effective forecasts are mostly based on their empirical systems developed for the observed order of "before and after" naturally existing in the evolutions of events, learned from first-hand experience. That is, in the field of meteorology, there is the problem of theories not agreeing with the practice. Based on the fact that all the theoretical researchers in meteorology from around the world cannot practically forecast weather, we have established the blown-up theory of evolution [OuYang, McNeil, & Lin, 2002; Wu & Lin, 2002] and the prediction method of digitized information [OuYang, Chen, & Lin, 2009; Lin & OuYang, 2010] by seriously considering the empirical experiences of the front-line forecasters. In other words, in the field of meteorology there is a major disagreement between theories and practically useful methodologies; and front-line forecasters in general do not recognize the validity of the system of conventional theories. (For the convenience of the reader and to make this presentation self-contained, let us briefly explain the two theories just mentioned here. First, the blown-up theory [Wu & Lin, 2002] is initially established by our joint effort based on OuYang's 50-years of practical success with his timely predictions of zero probability disastrous weathers. It is a systems theory focusing on whole evolutions instead of local, relatively static phenomena. What is found, along with many other important results, is that mathematical nonlinearity stands for eddy motions, which naturally make all the scientific tools based on the concepts of continuity and differentiability over Euclidean spaces generally invalid. Also, it is discovered that disastrous weather systems form in the discontinuous regions between eddy movements of the atmosphere. That explains in general why when one employs deterministic means to investigate evolutions, one most likely experiences chaos or stopped computers due to numerical overflows. Secondly, the prediction method of digitized information is developed by us to take advantage of the widely available data, especially those that are automatically collected in meteorology, on our new understandings of the concept of time. By taking time out of the computation, and by converting nonstructural data into structures in relevant phase spaces, forthcoming changes in structures can be readily seen and predicted ahead of the appearance of disastrous weather systems, as seen in [Lin & OuYang, 2010].)

About twenty years after Lorenz's chaos doctrine was initially published, the fact that out of the "deterministic" equations of the chaos doctrine one produces the "indeterminate" chaos was warmly embraced by the then-modern community of theoretical physics, which was then totally engulfed in the heat wave that "God plays dice," leading to the overheated discussions of chaos theory. This theory was even seen as epoch-making in theoretical physics throughout the community of physics. At the very moment when the chaos theory was seen as a raging wild fire, to satisfy our students' demands due to our successes in predicting (near-) zero-probability disastrous weather systems and publicized disagreement with the conclusions of chaos theory, we organized research funds and man-power to take a closer look at the problem of nonlinearity, Lorenz's work related to chaos, and his followers' works by employing various computational schemes on their well-discussed models [OuYang & Lin, 1997; OuYang, 1998; Lin *et al.*, 2001]. In order not to make this presentation too long, in the following we will only focus on the problems existing in the initial model that Lorenz himself proposed and how he computationally obtained the phenomenon of chaos. When Lorenz proposed his chaos theory, he did not indicate that nonlinearity stands for the following problem of rotationality [OuYang, McNeil, & Lin, 2002; OuYang, Chen, & Lin, 2009; Wu & Lin, 2002; Lin & OuYang, 2010]:

**Problem 1.2.** *It is shown in [Lin, 1998; Lin, 2008; Lin & OuYang, 2010] that, in terms of physics, mathematical nonlinearity stands for eddy motions in curvature spaces. Then, how can anyone study eddy motions and their interactions accurately by only employing the linear theories developed in the system of numbers and by limiting oneself only within Euclidean spaces made up of linear axes or dimensions?*

In the past hundred some years various propositions of randomness have been derived out of the calculus of mathematical logic; and the system of classical mechanics cannot fathom irregular events. So, it is natural for people to think about employing stochastic systems for them to understand matters that seem to be by chance or random. Additionally, Lorenz's chaos theory stems from some nonlinear equations of the classical system of dynamics; and in the past one hundred years due to the employment of mathematical logic, the classical system of dynamics had been considered as the fatalism of deterministicity. Therefore, it seems that by using the "indeterminacy," produced out of the "deterministic" chaos theory, one can completely replace the Newtonian realm by that of a "probabilistic universe".

## 1.2. Lorenz's Modeling and Problems of the Model

The form of Lorenz's model comes from models of physics. However, in essence it is a mathematical model established by using parametric dimensions. In other words, the model itself does not comply with the underlying physical meanings. In particular, Lorenz's model [1963] is established on the basis of Saltzmann's [1962] convection model of a small scale atmosphere. However, Saltzmann's model is only a simplified small scale $X - Z$ spatial convection model, suffering from the major problem of assuming the atmospheric density to be constant. For problems about weather changes, even for large scales one cannot assume that the atmospheric density is a constant. Evidently, if the atmospheric density is not a constant, Saltzmann's convection model does not hold true. This is because when the atmospheric density is not constant, Saltzmann's convection model becomes a problem of turbulences of rotational movements under the effect of stirring forces [Wu & Lin, 2002] so that it can no longer be established using the means of the first push. When we initially started to investigate Lorenz's model, we chose not to participate in the analysis of the so-called problem of chaos theory by pointing out the fact that Saltzmann's convection model does not contain any practical significance and the consequent model as modified by Lorenz is even further away from the reality. However, in the early 1980s the community of physics vigorously embraced chaos theory emphasizing Lorenz's model, which later spilled over to the area of meteorology. This historical fact indicates that even in the 1980s, the scientific community still did not seem to accept that mathematical models cannot be "simplified" to such a degree that the underlying physical meanings are totally lost. That is, the Saltzmann's [1962] convection model is the following:

$$\begin{cases} u_t + uu_x + wu_z = -\frac{1}{\rho_0}p_x + v\Delta u \\ w_t + uw_x + ww_z = -\frac{1}{\rho_0}p_z - g\pi + v\Delta w \\ \pi_t + u\pi_x + w\pi_z = \frac{N^2}{g}w + K\Delta\pi \end{cases} \tag{1.1}$$

where $u$ and $w$ are respectively the speed components along the $X$ and $Z$ directions, $\pi = \frac{\rho}{\rho_0}$, $\rho_0$ is a constant, $N^2 = \frac{g}{\theta}\frac{\partial^2\theta}{\partial x^2} + \frac{\partial^2\theta}{\partial z^2}$ is the stratification parameter, and $v$ and $K$, respectively, are the viscosity and heat-conduction coefficients. All the subscripts stand for the corresponding partial derivatives with $\Delta = \frac{\partial^2}{\partial x^2} + \frac{\partial^2}{\partial z^2}$ being the Laplace operator. Evidently, Eq. (1.1) is a mathematical equation of the first-push system of classical mechanics. Because $\rho_0$ is a constant, its acting force is merely a push. This end implies that Saltzmann's [1962] atmospheric convection model does not comply with the concept of a stirring force of the solenoidal term of the circulation theorem of the meteorological science [OuYang, McNeil, & Lin, 2002; Wu & Lin, 2002]. That is, the left-hand side of Eq. (1.1) represents the nonlinearity of the underlying movement, which should correspond to the nonlinearity of the stirring force of the solenoidal term. In other words, Eq. (1.1) is already a mathematical model that does not agree with physical reality, where the atmospheric density changes drastically in the vertical direction.

As is well known, the problem of severe weather convections has been a difficult issue for the front-line meteorologists [Lin & OuYang, 2010]; and it is in the vertical direction that severe convections possess their strong characteristics. In Eq. (1.1), $N^2$ involves the problem of vertical stratification of the atmosphere. For convective weather, $N^2$ cannot be treated as constant. When attempting to resolve the problem of convective weather, one has to pay attention to the changes of the atmosphere in the vertical direction [OuYang, McNeil, & Lin, 2002; OuYang, 1998].

No doubt, the ultimate goal of scientific research is to describe physical reality and to develop means for resolving practical problems. However, model (1.1) itself contains the problem of inconsistency. Although Lorenz refers to Eq. (1.1) as a nonlinear model, in terms of the fundamental concept of nonlinear models, Eq. (1.1) does not represent a feasible nonlinear model. Even though it is mathematically classified as a weakly nonlinear model, that does not mean that such a mathematical term can truly represent reality. Due to the rotationality explicitly contained in nonlinearity [OuYang, McNeil, & Lin, 2002; OuYang, Chen, & Lin, 2009], one has to consider what Lorenz did as mathematical play.

Lorenz further introduces a flow function such that

$$u = \Psi_z \text{ and } w = -\Psi_x \tag{1.2}$$

Substituting Eq. (1.2) into Eq. (1.1) leads to

$$\begin{cases} (\Delta\Psi)_t + J(\Psi, \Delta\Psi) = -g\pi_x + v\Delta^2\Psi \\ \pi_t + J(\Psi, \pi) = \frac{N^2}{g}\Psi_x + K\Delta\pi \end{cases} \tag{1.3}$$

where $J(,)$ stands for the Jacobian, and all other symbols are the same as in Eq. (1.1). Although similar situations have been treated in such a fashion in classical mechanics, that does not mean that doing so is always reasonable and correct. Based on the basic concepts of the circulation theorem, rotational movements come from eddy sources. That is, the left-hand side of the first equation in Eq. (1.3) stands for a rotational vorticity, which cannot correspond to the push of a gradient force of the right-hand side. Therefore, the essence of Eq. (1.3) is different from that of Eq. (1.1). Even if we did not know the rotationality naturally contained in nonlinearity, we would still have to realize that pushes could only correspond to elastic effects instead of eddy effects. Therefore, in the conventional research of mechanics, there has been a problem of confusing material rotationality with material irrotationality in the name of quantitative formality. Here, by quantitative formality, we mean the formality of quantities (or numbers) that can be employed uniformly across the board in all scientific investigations with specifics ignored. For instance, behind each collection of objects, there is a count (a number) of how many of the objects there are. However, mathematical manipulations of these counting numbers do not necessarily reflect the physical properties of the objects involved.

By following the widely employed method of spectral expansions on the flow function and the non-dimensionalized density $\pi$, Lorenz obtains

$$\begin{cases} \Psi = \sqrt{2}\frac{K(n^2+k^2)}{kn} X \sin kx \sin nz \\ \pi = \sqrt{2}\frac{(n^2+k^2)}{gk^2 n} Y \cos kx \sin nz - Z\frac{(n^2+k^2)^3}{gk^2 n} \sin 2nz \end{cases} \tag{1.4}$$

where $k$ and $n$ stand for the wave numbers along the $X$- and $Z$-directions, $X$, $Y$, and $Z$ the coefficients of the spectral expansions, respectively representing the intensity of the movement, the strength of the linear stratification, and the strength of the nonlinear stratification. To this end, there is no doubt that it involves the problem of satisfying the required conditions of using the spectral expansions. It is because for successfully employing spectral expansions, one has to first guarantee the functions to be expanded have continuous first-order and second-order derivatives. Now, since the left-hand side of the first equation in Eq. (1.3) stands for a problem of rotationality of the nonlinearity, and rotations possess such a duality that the original function representing the interaction between the co-existent convergent and divergent spins are not guaranteed to be continuous. Therefore, the required guarantee of continuity of the first and second order derivatives is lacking.

Next, the central problem of spectral expansions is that the orthogonality used in Fourier analysis can eliminate quantitative singularities, where a quantitative singularity stands for a moment for the underlying quantitative variable to approach infinity. However, singularities themselves are the exact mathematical characteristics of Eq. (1.3). When these singularities are eliminated, the mathematical properties of the original model will have to be altered. With this understanding, one can see that Eq. (1.4) not only cannot represent Eq. (1.3), but also represents a problem of falsification. In other words, the mathematical models in Eqs. (1.3) and (1.4) have nothing to do with each other so that the consequent works no longer mean anything relevant.

Thirdly, Lorenz's parametric substitution of stratification can be said to be strange and unexplainable as either a mathematical or physics problem. In particular, it involves the following problems:

**(1)** The physical quantity $N^2$ of the stratification parameters is a constant in both Eqs. (1.1) and (1.3). However, in the expanded form in Eq. (1.4), this constant is split into two variables $Y$ and $Z$, respectively representing the strengths of the linear and nonlinear stratification. Because in general, linear and nonlinear effects cannot cancel each other, in Eq. (1.4), the constant $N^2$ becomes a variable so that the model in Eq. (1.3) is not equivalent to that in Eq. (1.4).

**(2)** The originally constant physical quantity of the stratification parameter becomes two variable physical quantities $Y$ and $Z$. Even according to the traditional methods of treatment, what is done here cannot be easily explained in terms of physics and can be seen as a misuse of mathematical parametric dimensions. In order to distinguish between linearity and nonlinearity, one can surely apply the first-order term and higher-order term of a certain physical quantity instead of employing two physical quantities. In terms of physics, using two variable physical quantities to describe one constant physical value can be seen as a work of quantification unique to Lorenz. Such a desire to employ quantification is rare in the history of modern science. No doubt, for such a paper to generate so much interest, there must be some correspondingly very important reasons. Note: Among many reasons is the desire of the community of physicists to pursue the concept of a probabilistic universe.

**(3)** Evidently, manipulations of mathematical physics equations are made possible by employing the technique of non-dimensionalization of physical quantities. What is worth noticing is that the symbol $\pi$ in the second equation of Eq. (1.4) is a non-dimensional quantity. Thus the right-hand side of the equation should also be non-dimensional. However, the variable $g$ does have dimensionality, and so do the corresponding stratification quantities $Y$ and $Z$. No doubt, the dimensionality of the same physical quantity has to be the same. However, the physical quantity of stratification is respectively represented by $Y$ and $Z$ linearly and nonlinearly so that different dimensions have to be introduced due to the differences in the involved linearity and nonlinearity. Therefore, there is a formal logical contradiction in Lorenz's non-dimensionalization. (In other words, by analyzing the units involved, the second equation in Eq. (1.4) cannot hold true). Evidently, when one substitutes a constant physical quantity by using two variable physical quantities for representing respectively linear and nonlinear stratifications, one must ignore a few formal logic problems. For a more detailed discussion of non-dimensionality, see [Lin & OuYang, 2010, pp. 89–93].

Since Lorenz comes from the community of meteorology, he must have already known that nonlinearity implies solenoidal effects as pointed out by V. Bjerknes's circulation theorem of the meteorological science, causing the appearance of rotational eddy currents in the atmosphere and oceans. What is ironic is that the original model (Eq. (1.1)) of Lorenz's work is proposed by Saltzmann in 1962. That is, neither Saltzmann nor Lorenz has followed the circulation theorem, and both of them have ignored the essential differences between wave motions and eddy motions.

From the discussions given so far in the previous paragraphs, the reader should see that in order to employ available mathematical techniques in the study of modern science, some of the contemporary scholars have pushed beyond the accepted boundaries. This end also explains why in the system of modern science all disturbances are treated as wave motions so that available mathematical techniques can be "beautifully" employed without considering the consistency of formal logic of the mathematics involved while ignoring the essential problem of disregarding the fundamental underlying events. This realization also clearly suggests that the physics community was interested in chaos theory more as a way of explaining of a probabilistic universe than of understanding what physical chaos stands for. Lorenz substitutes Eq. (1.4) into Eq. (1.1) and lets $\tau = \pi^2 h^{-2}\left(1+\alpha^2\right) Kt$ be $t$ to produce the following mathematical model of his chaos theory:

$$\begin{cases} X' = -\sigma X + \sigma Y \\ Y' = rX - Y - XZ \\ Z' = XY - bZ \end{cases} \tag{1.5}$$

where $A' = \frac{dA}{d\tau}$ $(A = X, Y, Z)$, $\sigma = vk^{-1}$ is the Prandtl number, and $r = \frac{R_\alpha}{R_c}$ is the Rayleigh number with $R_\alpha = g\alpha h^3 \Delta T v^{-1} K^{-1}$, $R_c = \pi^4 \alpha^{-2}\left(1+\alpha^2\right)^3$ and the threshold values $R_{cm} = \frac{27\pi^4}{4}$ and $b = 4\left(1+\alpha^2\right)^{-1}$. Here $\alpha$ stands for the coefficient of heat expansion, $h$ the plane distance, $\Delta T$ the temperature difference, and all other symbols are the same as before. In the computations of Lorenz's chaos theory, $b = \frac{8}{3}$, $\sigma = 10$, and $r \approx 1 \to 233.5$. That is, the $r$-value can be varied. Equation (1.5) is the first mathematical model of chaos in Lorenz's chaos theory. Its limitations as a physical model can be summarized as follows:

(1) Because Saltzmann's atmospheric convection model in Eq. (1.1) describes the form of fluid movement using partial derivatives, it is referred to as a model written in the Euler language. If total differentials are employed, then the resultant model can no longer be seen as an expression in the Euler language. For instance, if the total differentials are employed, the first equation in Eq. (1.1) would become

$$\frac{du}{dt} = u_t + u \cdot u_x + v \cdot u_y + w \cdot u_z = -\frac{1}{\rho_0} p_x + v\Delta u$$

which is known as an expression in the Lagrange language. However, when compared to the first equation in Eq. (1.1), this expression has an additional term representing the component in the $Y$-direction. That is why Saltzmann's model in Eq. (1.1) is seen as an Euler model of a simplified N-S equation in the Euler language. And in the process of simplification, the parameter $\rho_0$ is introduced and $v$ is fixed to be constant.

Also, the linear form of the dissipative term in the N-S equation itself suffers from the problem of whether it agrees with the underlying physical reality. The details are omitted here except we like to point out that any realistic dissipative process is an effect of rotational eddy currents so that the corresponding mathematical model has to be nonlinear. That is, there is a need to reconsider the problem of mathematical modeling in classical fluid mechanics developed since the N-S equation was initially established.

(2) Evidently, when the atmospheric density in Eq. (1.1) is taken as a constant ($\rho = \rho_0$), the mathematical properties of the N-S equation are altered so that the physical significance of the external forcing term is distorted. That is, when it is assumed that $\rho = \rho_0$, we should not think the equation is "simplified." Instead, what is done is to replace the original nonlinear source of stir by a fabricated source of push. That is why both Saltzmann's Eq. (1.1) and Lorenz's Eq. (1.3) have continued the tradition without confronting the essential difference between nonlinearity and linearity.

(3) The familiar method of spectral expansion is generally used to materialize quantitative stability by eliminating quantitative singularities (by making use of the orthogonality of Fourier expansions). However, quantitative instability and singularity are the fundamental characteristics of nonlinearity.

In terms of quantitative analysis, all possible changes are

$$\frac{dx}{dt} = 0, \frac{dx}{dt} \to \infty, \frac{dx}{dt} \to 0$$

where $x$ stands for a quantity with time $t$ temporarily being the Newtonian numerical time. For a discussion of what time is, see [OuYang, McNeil, & Lin, 2002; OuYang, Chen, & Lin, 2009; Lin, 2009]. Let us now look at these three cases one by one:

**(1)** $\frac{dx}{dt} = 0$ stands for quantitative invariability under the concept of stability. It can also be seen as such a problem that summarizes the essence of modern science, and becomes naturally the initial value stability (or known as an initial-value automorphism) of the well-posedness conditions of the modern mathematical physics equations. It is exactly so that it reveals the essence of how the system of classical mechanics ignores changes in materials [OuYang, McNeil, & Lin, 2002; Bergson, 1963; Koyré, 1968; Prigogine, 1980].

**(2)** The second scenario $\frac{dx}{dt} \to \infty$ corresponds to the concept of quantitative instability, also known as either quantitative unboundedness or quantitative singularity. However, singularities correspond very well with transitionalities in curvature spaces [OuYang, McNeil, & Lin, 2002; Lin 2009]. Thus eliminating quantitative singularities is getting rid of transitionalities in evolutions. In other words, it is exactly because of the quantitative singularities that mathematical properties of nonlinearity and the transitionalities existing in the evolutions of events are revealed [OuYang, McNeil, & Lin, 2002; Lin, 2009].

**(3)** The third scenario $\frac{dx}{dt} \to 0$ does not have any substantial difference from scenario (1). It only emphasizes the process and existence of such mathematical equations that cannot be accurately computed due to the involvement of quasi-equal quantities [OuYang, McNeil, & Lin, 2002]. The inaccurate computations of quasi-equal quantities are vividly embodied in Lorenz's chaos.

## 1.3. Computational Schemes and What Lorenz's Chaos Is

The core of computational schemes is about how to deal with quantitative instability. To this end, [OuYang, McNeil, & Lin, 2002; Lin *et al.*, 2001] have done comparative analysis on various computational schemes, uncovering the fact that nonlinear quantitative instabilities are about explosive quantitative growths and that the results of the unstable direct difference scheme are much closer to the analytic solutions than those produced by other computational schemes. To this end, let us first rewrite Eq. (1.5) using the direct difference scheme as

$$\begin{cases} X_{n+1} = X_n + (-\sigma X_n + \sigma Y_n)\,\Delta\tau \\ Y_{n+1} = Y_n + (rX_n + Y_n - X_n Z_n)\,\Delta\tau \\ Z_{n+1} = Z_n\,(X_n Y_n - bZ_n)\,\Delta\tau \end{cases} \tag{1.6}$$

where $\Delta\tau$ stands for the length of the time steps used in the integral over time, and the subscripts $n$ and $n+1$ represent the variables' values at the time moments $n$ and $n+1$, respectively.

Evidently, either Eq. (1.5) or (1.6) is a system of mathematical equations in three unknowns, containing both linear and nonlinear terms. Those familiar with numerical computations will see that Eq. (1.6) represents a trajectory of reciprocating double spirals in the phase space.

By continuing the terminology of the well-posedness of the traditional mathematical physics equations, Lorenz particularly emphasizes the initial-value sensitivity and considers Eq. (1.6) as an example of such. However, computation of nonlinear products of quantities (with non-small effective values) is sensitive not only to the initial values, but also to any of the variables involved, including time steps and parameters.

If in Eq. (1.6) we let the initial value be $\tau \approx 0$ and take $X = 3$, $Y = 2$, and $Z = 4$, then we have the following:

(1) The integral results using non-smoothing schemes with relatively small time steps: If $\tau = 0.02$, then the computational outcomes of the direct difference scheme are given in Fig. 1.1. What should be noted is that

the direct difference is an unstable scheme in traditional computational mathematics. However, when the time steps are relatively small, one can also obtain the same results as those produced by smoothing schemes.

Note: One of the reasons why we stopped the computation in Fig. 1.1 at the step of $1 \times 10^7$ was that at the time we did the initial calculation, the speed of the computer was quite slow. A second reason was our desire to emphasize the fact that the so-called sensitivity is not merely limited to the initial values; instead it was also sensitive to the time step . And when the time step is small, it is equivalent to making the initial values used in the computations small, leading to even smaller products of these small values. This end is similar to the essence that the square of 0.1 is a much smaller number than the starting number. On the contrary, nonlinear products of non-small numbers give rise to much greater numbers, leading to numerical instability in the calculation. Our purpose of providing this specific example is to show that even with one of the most unstable schemes, we can still produce stable computational results. This conclusion suggests that Lorenz's so-called *butterfly* is not from the instability of nonlinearity, but rather is a result of errors in the large number of infinitesimal differences. For more details, please see the appendix of this paper or [Lin, 1998; Wu & Lin, 2002; Lin & OuYang, 2010].

Based on what is discussed here, one can see that when dealing with any problem of small quantities using small time steps, no matter whether we use stable or unstable computational schemes, we can produce the "butterfly" effect. For the purpose of comparison, we applied the same initial values and time steps as used in Fig. 1.1 by employing the smoothing Runge-Kutta scheme. The results are given in Fig. 1.2.

The computational procedures indicate that both of these calculation schemes do not suffer from memory spills and can be continued with an increasing trajectory density. By directly comparing Figs. 1.1 and 1.2, it is easy to see that both computational schemes produce roughly the same results without any substantial difference except that the graphs in Fig. 1.2 are smoother than those in Fig. 1.1 and the left-hand spiral is a bit smaller. What should be pointed out is that the results in Fig. 1.1 are obtained by using an extremely unstable computational scheme, where as long as the time step is sufficiently small, the same outcomes as what Lorenz obtained by using smoothing scheme can be produced. Although at the time we initially did our computation, the computer was very slow and required three days, the stable computation could be continued without any sign of potential interruption. So, when we constructed the graphs in Fig. 1.1,

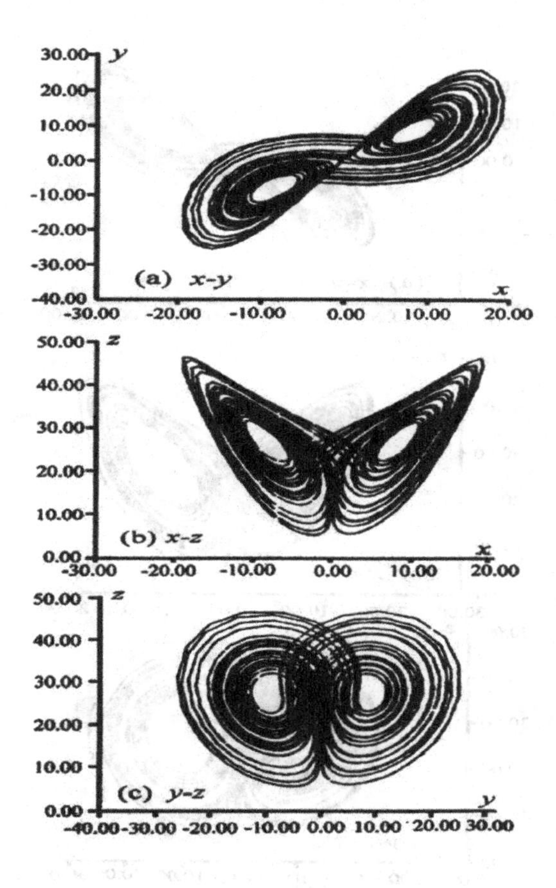


Fig. 1.1. Lorenz model with small time steps and a non-smoothing scheme.

we stopped our computation at the step of $1 \times 10^7$. On the contrary, in the computation of nonlinearity, one faces either explosive instability or the stability of drastic absolute value decrease where the rate of nonlinear decrease of small values is much faster than that of the linear drop.

Evidently, the previous numerical experiments imply that when the time step is relatively small, both smoothing and non-smoothing schemes are essentially the same. Thus a natural question arises: Since the direct difference scheme is extremely unstable, how can it produce essentially the same results as a stable smoothing scheme?

To address this problem, we followed an on-screen analysis of the computational procedure. Our results reveal that the computational stability

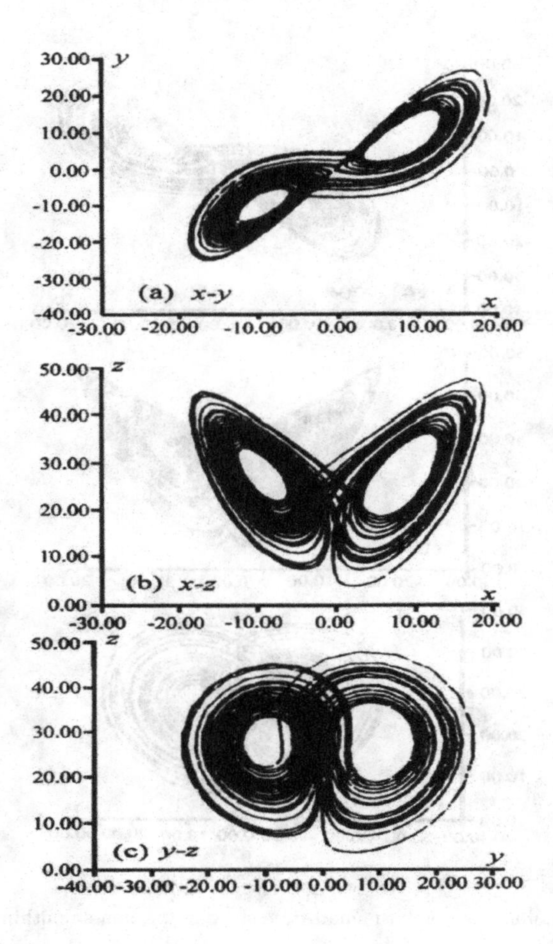


Fig. 1.2. Lorenz model with small time steps and a smoothing scheme.

of nonlinearity has nothing to do with the schemes employed; instead, it has everything to do with whether or not the computational values fall into non-first-order product(s) of small quantities and infinitesimal difference(s) of large quantities, meaning the differences of quasi-equal quantities. Because the values that actually participated in the computation procedures are minuscule error values, that is how the two-leaf spiral trajectories in Figs. 1.1 and 1.2 are created. And the similarity of these two-leaf spirals to the so-called butterflies is completely caused by the reciprocations of the remnants of the errors in the phase space of Lorenz's model.

Note: Based on our repeated computations and comparisons of the computational outcomes, we conclude that nonlinear products of non-small quantities are the essence of the numerical instabilities involved, which behave as explosive quantitative growth; and the nonlinearity of small values is shown as the extreme stability of "explosive decay," which is much faster than linear decay. For more details, see the appendix of this paper, or [Lin, 1998; Wu & Lin, 2002; Lin & OuYang, 2010].

Speaking in nontechnical terms, the so-called butterfly effect of chaos theory can be compared to Shakespeare's Venice merchants who could never cut off one pound of meat exactly. Or, as the ancient Chinese calculator Zhan Yi said, there are things that numbers cannot describe. However, the "butterfly"Lorenz has seen is a characteristic of nonlinearity.

So, there is no doubt that small differences of large quantities or the inaccurate equations involving quasi-equal variables become incalculable problems or equations due to the inevitable error-value computations. In other words, Lorenz's chaos theory comes from the problem of computational inaccuracy of equations involving quasi-equal quantities.

(2) The computational results with enlarged time steps:

By increasing the time steps, changing the initial values, or altering the parameters, one can make all the consequent quantitative computations avoid the situation of infinitesimal differences of quasi-equal large values. Let us now provide the corresponding computational results obtained by increasing the time step. At this junction, what needs to be remembered is the following fundamental characteristics of nonlinear quantitative calculation: Non-first-order products of small values behave as a drastic decrease; while non-first-order products of non-small values become an explosive increase. Only the latter constitutes the quantitative instability of nonlinearity.

If we take the time step $\Delta\tau = 0.0651317\bar{8}$, Fig. 1.3 shows the computational outputs of the direct difference scheme. Fig. 1.4 reveals the obvious characteristics of smoothing schemes, where the two-leaf spirals are no longer smooth, leaving behind a track of forced smoothing alteration in the direction of increasing values. All of the computational values in the three phase planes in Fig. 1.3 interrupted the computation due to spills in the negative $Y$ and $Z$ directions. In other words, as soon as the computation reaches a certain number $n$ of steps (in this example, $n = 18$), Lorenz's chaos (the two-leaf spiral butterfly effect) disappears.

Evidently, for nonlinear problems, quantitative instability is the fundamental characteristic of nonlinearity. Or said differently, nonlinear mathematical models possess the characteristics of escaping the continuity of

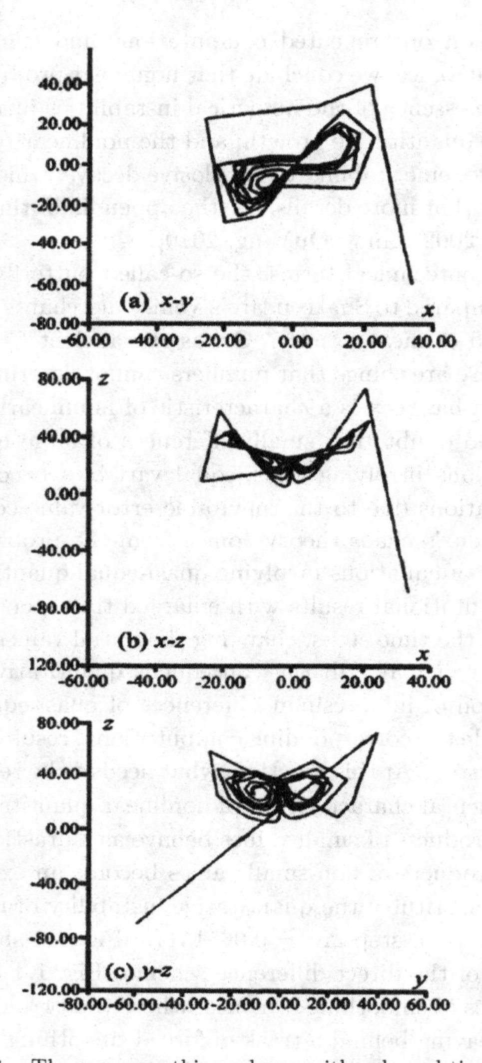


Fig. 1.3. The non-smoothing scheme with enlarged time steps.

Euclidean spaces and entering the transitionality of non-Euclidean spaces, which stands for a change in movement, such as direction.

At this point, we mention that even if we take $\Delta\tau = 0.0651317\bar{7}$, the consequent computational results will produce error spirals (or Lorenz's butterfly effect) similar to that of Fig. 1.1, but if the time step is increased by a tiny amount of $0.0000000\bar{1}$, instability onsets, illustrating the problem

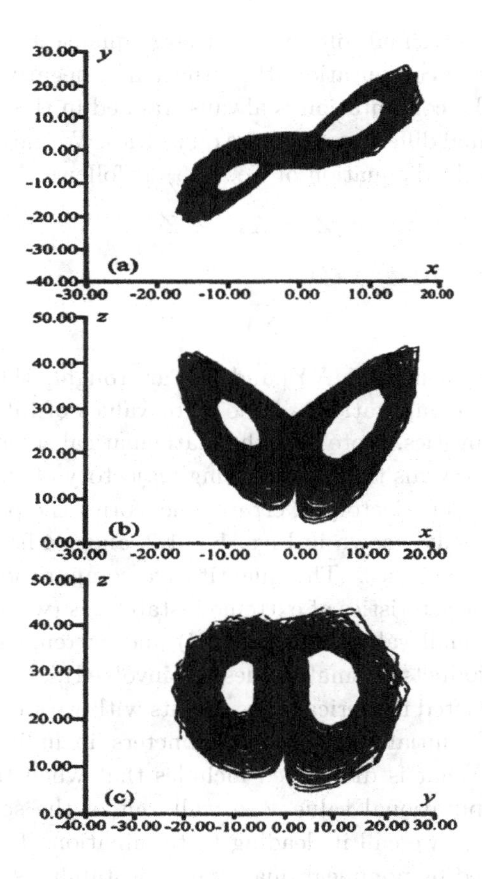


Fig. 1.4. The smoothing scheme with enlarged time steps.

of error value computations involving infinitesimal differences of large quantities.

Fig.1 .4 shows the computational results of the smoothing Runge-Kutta scheme under the same conditions as in Fig. 1.3. Its characteristics are similar to those "butterfly" error effects of Figs. 1.1 and 1.2. Similar results can be obtained by using the smoothing scheme of "one step forward and three steps backward."

Our on-screen tacking reveals that because of the computational explosive growth along the negative $Z$-direction, the spill causes the computer to stop working so that the calculation is interrupted. That is the only reason why the computational results in Fig. 1.3 escape the "butterfly" error effect of infinitesimal differences of large quantities. For Fig. 1.4, in the entire process of computation, the situation of negative $Z$-values never appears so that the computation is always trapped in the "butterfly" error effect of infinitesimal differences of large quantities. We can see this problem directly from the third equation of Eq. (1.5) as follows: If $Z > 0$, we have

$$Z' = XY - bZ \tag{1.7}$$

If $Z < 0$, we have

$$Z' = XY + b|Z| \tag{1.8}$$

Evidently, the magnitudes of $|XY|$ and $|bZ|$ are roughly the same, and thus Eq. (1.7) involves computations of the error values of infinitesimal differences of large quantities. Note that the mathematical structure of Lorenz's model (1.5) itself reveals the reciprocating trajectory of the computed values. That is why its trajectory of error values forms the pattern of a "butterfly"Thus the results shown in Figs. 1.1, 1.2, and 1.4 have nothing to do with nonlinear instabilities. The quantitative computations of nonlinearity possess the characteristics of extreme instabilities (when non-first-order products of non-small values are involved) and extreme stabilities (when non-first-order products of small values are involved).

Also, we conducted numerical experiments with various time steps (and numbers of steps), initial values, and parameters, including those of adaptive time steps. What is discovered includes that when the time step increases, the computational values can walk out of the so-called chaos or the error spirals, as we call it, leading to termination of the computation due to spills caused by nonlinear quantitative instabilities. If the time step decreases, then the computational values can exhibit error spirals. Also, just before a spill appears, if the computational scheme is changed to a smoothing scheme or the parameters are reduced in magnitude, then the computational results once again return to the error spirals. And by alternatively employing weak smoothing schemes, such as the second order Runge-Kutta scheme, or large time steps, large parameters, large initial values, and other strong smoothing schemes, we can make the computational values either escape or enter error spirals [Lin *et al.*, 2001] at will.

(3) The computational results when the initial values are negative:

Let $\tau = 0$, the initial values $X = -2, Y = -2, Z = -4$, and the parameter $b = \frac{8}{3}$. When $\Delta\tau \leq 0.05792709\dot{6}$, the computational results enter the error spiral with a graph, which is omitted, similar to Fig. 1.1. When $\Delta\tau \leq 0.05792709\dot{7}$, which is only $0.00000000\dot{1}$ greater than the previous

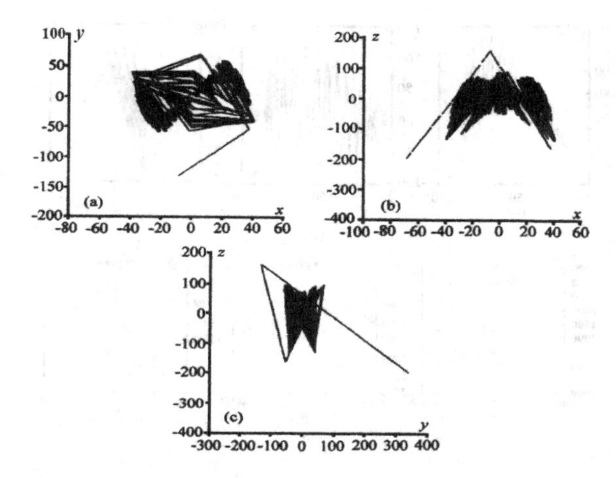


Fig. 1.5. The non-smoothing scheme applied on Lorenz's model with increasing parameter.

value, the computational results jump out of the error spiral right after they entered it; and soon after that the computation stops due to computational instability (the figure is omitted). However, if we change the scheme to a smoothing scheme before the spill, the computational values will once again enter into the error spiral (the figure is omitted).

(4) The computational results when the parameters are varied:

Scheme 1: To make the comparison easier, we take the same initial values as in (3) and let $\Delta\tau = 0.05$. That is, this time step is smaller than that used in (3) where the computation enters an error spiral. Our purpose of doing so is to make sure that when a parameter is varied, no quantitative instability will be created by the choice of a too large time step. Let us now vary the parameter $b$ in two situations. When $b$ falls between $\frac{8}{3}$ and 331849, the computational results are trapped in an error spiral. The figure is omitted. When $b \geq 331850$, as soon as the calculation reaches the $97914^{th}$ step, the computational results suddenly escape the error spiral and the calculation stops due to a spill in the form of $Z < 0$, see Fig. 1.5 for more details. What is interesting is that the pattern in the $X \times Z$ phase space looks like a small flying bird with its wings wide spread out instead of a "butterfly" if we stop the calculation at the $97912^{nd}$ step. If Lorenz had examined the sensitivity to parameters he would have seen a bird fluttering its wings instead of a butterfly. What should be pointed out is that in order to show the difference between being trapped in error-value computations

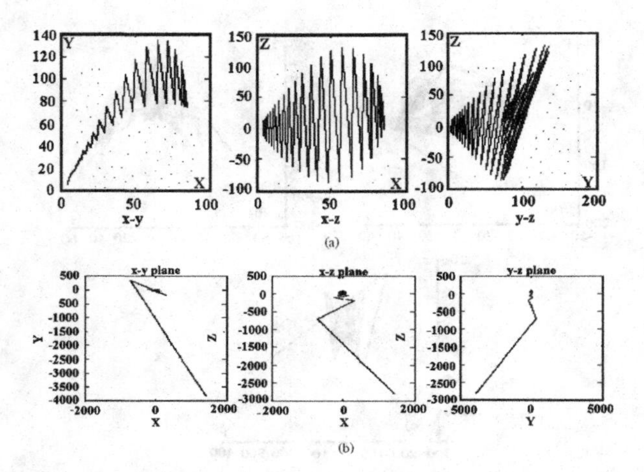


Fig. 1.6. (a) $\sigma = 10$, $r = 28$, $b \in [198, 241]$, $\Delta\tau = 0.008$, $X = 5$, $Y = 3$, $Z = 9$. (b) $\sigma = 10$, $r = 28$, $b > 249.2309892$, $\Delta\tau = 0.008$, $X = 5$, $Y = 3$, $Z = 9$, $n = 69751$.

and the quantitative unstable walk out of the error-value computations, when constructing Fig. 1.5, we only kept the computational results of the steps between 96000 and 97914. Later our former student and current colleague Professor Yigang Chen repeated our work using newer computers.

Scheme 2: For the sake of comparison, let the initial values be positive, such as $X = 5, Y = 3$, and $Z = 9$, and the time step $\Delta\tau = 0.008$. That is, for this time step all the previous computations can enter the error spiral. Similar to Scheme 1 above, we also vary the parameter $b$ in two cases. When $b$ is between 198 and 249.23, the computational results present a pattern like a smoothly opened or slightly curved tree leaf as shown in Fig. 1.6(a). For $b = 249.2309892$, when the calculation reaches the $69750^{th}$ step, the results escape the error spiral and the calculation stops with $Z < 0$ as shown in Fig.1. 6(b). For the convenience of generating the graph, Fig. 1.6(b) only includes the results of the step 69751.

(5) Varying the Parameter $r$:

When $r$ is between 28 and $86.7217888351\dot{3}$, the computational results are trapped in an error spiral as shown in Fig. 1.7(a). The first pattern resembles that of the *ancient Chinese Taiji graph*, and the second looks like a *flying butterfly* viewed from the front or the back. When $r = 86.7217888351\dot{4}$, the results suddenly jump out of the error spiral at step 792 and the calculation stops due to a spill in the form of $Z < 0$ (Fig. 1.7(b)). Thus the parameter $r$ is very sensitive, reaching the level of 1%.

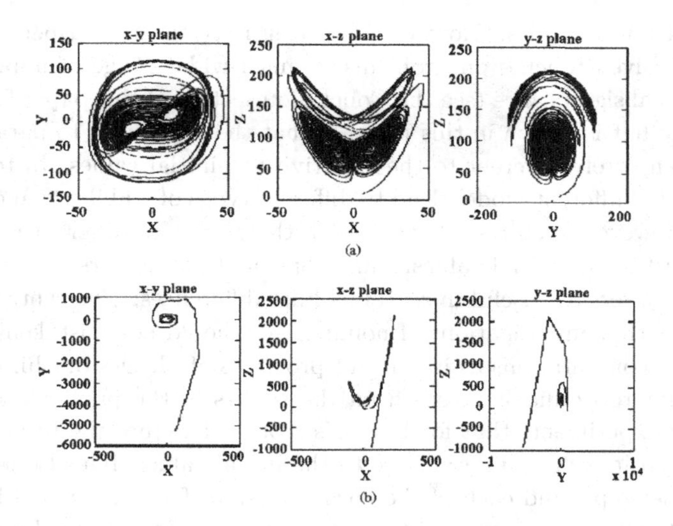


Fig. 1.7. (a) $r = 86.72178883513$, $b = 28$, $\sigma = 10$, $n = 80000$, $\Delta\tau = 0.008$, $X = 5$, $Y = 3$, $Z = 9$. (b) $r = 86.72178883514$, $b = 28$, $\sigma = 10$, $n = 792$, $\Delta\tau = 0.008$, $X = 5$, $Y = 3$, $Z = 9$.

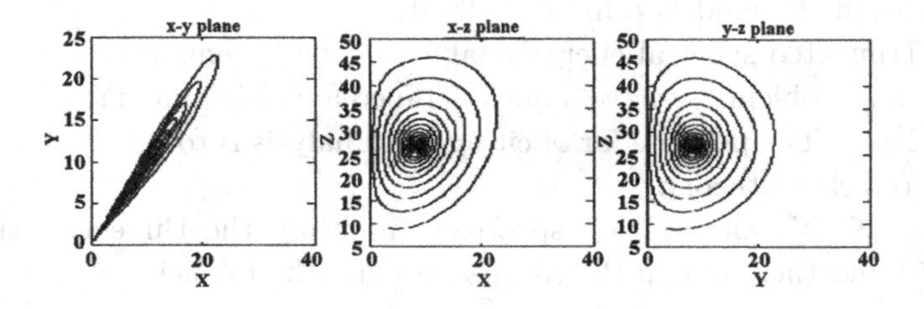


Fig. 1.8. $b = 28$, $\sigma \in [61.80311085, 233.1]$, $r = 28$, $\Delta\tau = 0.008$, $X = 5$, $Y = 3$, $Z = 9$.

(6) Varying the parameter $\sigma$:

When $\sigma$ takes values between 10 and 61.80311084, the computational results fall inside the error spiral with the patterns like a *feather pen* or the *age rings of a tree.* When $\sigma$ takes values between 61.80311085 and 233.1, the pattern looks like a *feather* as shown in Fig. 1.8. Note: First of all, we mention that the specific computations shown here were done by our former student Dr. Gongyi Chen on the current fast computers with a focus of testing the sensitivity to changes of parameters. The conclusions indicate that the so-called sensitivity of nonlinear quantitative computations exists also for either the magnitude of time steps or changes in parameters in addition to that of initial values. However, the sensitivity to other aspects of the models involved touches on problems of much wider range of importance and practical significance (see the Conclusion section of this paper for more details). What is shown in this paper is that the sensitivity to parameters can be completely different to the sensitivity to initial values. In terms of nonlinearity, different models lead to different types of stabilities and instabilities. However, even for a fixed model, there are also different scenarios with regard to the initial values, time steps, and parameters. To this end, faster computers are useful in revealing finer differences. The central problem here is that investigations of nonlinearity should not mistakenly treat error-value computations as important problems of physics or philosophy.

It is not too difficult to see from the results of the previous series of numerical experiments that for Lorenz's model, if we use the term "sensitivity," then it is not only sensitive to the initial values, but also sensitive to the time steps, and each of the parameters. In fact, the model is more sensitive to the parameters and time steps than to the initial values. With only about 1% of changes in the parameters or time steps, the computational outcomes can be totally different. Even for linear problems, similar results can be obtained except for different forms of presentation due to differences in the models [Lin *et al.*, 2001].

(7) Truncated spectral energy analysis of the Lorenz model:

For this problem, we have done a detailed analysis in [OuYang, McNeil, & Lin, 2002]. The purpose for of citing this analysis is to reveal the essence of Lorenz's chaos theory.

Using $X$, $Y$, and $Z$ to respectively multiply the three equations in Eq. (1.5) and then adding the results provides the following:

$$\frac{1}{2}\frac{d}{d\tau}\left(X^2+Y^2+Z^2\right)=-\left(\sigma X^2+Y^2+bZ^2\right)-b\left(\sigma+\gamma\right)Z \tag{1.9}$$

Let $E=\frac{1}{2}\left(X^2+Y^2+Z^2\right)$. Then when $E$ is a constant or when $\frac{dE}{d\tau}=0$, the truncated spectral energy is invariant because when $b,\sigma,r\neq 0$, one must have $X,Y,Z=0$. When $\frac{dE}{d\tau}>0$, that is when the truncated spectral energy increases, we have:

$$\sigma X^2+Y^2+bZ^2+b\left(\sigma+\gamma\right)Z<0 \tag{1.10}$$

because $b,\sigma>0$, if $\sigma>|r|$, then $Z<0$ and

$$\left|b\left(\sigma+\gamma\right)Z\right|>\left|\sigma X^2+Y^2+bZ^2\right|$$

or

$$|Z| > \left| \frac{\sigma X^2 + Y^2 + bZ^2}{b(\sigma + \gamma)} \right| \tag{1.11}$$

Therefore, when $Z < 0$ and Eq. (1.11) is satisfied, the truncated spectral energy increases, which agrees well with the fact that in the numerical computations, the calculation stops in the form of $Z < 0$.

When $\frac{dE}{d\tau} < 0$, the truncated spectral energy decreases so that we have

$$\left(\sigma X^2 + Y^2 + bZ^2\right) + b(\sigma + \gamma) Z > 0 \tag{1.12}$$

or

$$|b(\sigma + \gamma) Z| < \left|\sigma X^2 + Y^2 + bZ^2\right| \tag{1.13}$$

Evidently, if $Z > 0$, then the evolutionary values fall into the error spiral due to the appearance of infinitesimal differences of large quantities so that their practical significance is lost.

If $|b(\sigma + \gamma) Z| < \left|\sigma X^2 + Y^2 + bZ^2\right|$, then $Z$ can be either greater than 0 or less than 0. When $Z > 0$, the situation is similar to what is discussed above. That is, the Lorenz's chaos is trapped in an error spiral. When $Z < 0$ and Eq. (1.11) is not satisfied, the energy decreases due to the constraint of $\frac{dE}{d\tau} < 0$. In other words, the condition $Z < 0$ under this circumstance can still constitute a case of infinitesimal differences of large quantities so that the computational outcomes are falling into an error spiral.

This truncated spectral energy analysis points out some essential problems existing in the chaos model and the consequent theory. We also did a whole series of computations for other widely employed classical chaos models, such as the Brussel Oscillator, forced Brussel Oscillator, and Rössler model [Lin *et al.*, 2001]. This work required over three years of time and effort and a great amount of manpower. To spare the reader, the details are omitted.

## 1.4. Discussion

Although modern science advocates that assumptions are inevitable, these assumptions should not be introduced without constraint. In his modeling, Lorenz substitutes a constant "atmospheric stratification" by using two variables, which cannot even be plausibly supported or explained. It is because if one uses $Y$ and $Y^n$ to be the linear and nonlinear "atmospheric stratifications," then the model will be a two-dimensional parametric model, from which the butterfly form cannot be produced.

Our truncated spectral energy analysis produces the same results as the numerical experiments and reveals more essentially the problems explicitly contained in Lorenz's model and his chaos theory. That is, Lorenz's chaos is not produced from increasing truncated spectral energy; instead, it is a result of decreasing energy. Evidently, when the change in the energy decreases ($\frac{dE}{d\tau} < 0$), it means that the level of energy approaches invariability. That is, $E \approx$ a constant. That is also where the attractors come from.

Any realistic chaos or "orderlessness" means $\frac{dE}{d\tau} > 0$, which can truly reflect the explosive growth caused by nonlinear quantitative instabilities. The so-called *initial-value sensitivity* is not simply the "butterfly fluttering its wings" of the initial values' error spiral. The sensitivity also comes from the quantitative instability in the nonlinear effective values. Thus Lorenz's chaos is not the same as physical irregularities, or orderlessness, or the chaos of nonlinear growth. Rather it is the error-value computations of quantitative quasi-equalities. Lorenz employed his chaos to reveal the "mess" of differences, leading to the consequence of quasi-equality of such "twins" that are different only with a few body hairs. This end could be seen as analogous to the one pound meat of Shakespeare's Venice merchants, which could never be accurately cut.

Now, a significant question arises:

**Problem 1.3.** *How did a chaos theory in which both the model and numerical computations suffer from serious flaws ever become so influential in the physics of the 20th century?*

The relevant background is that mathematical logic has conducted the appearance of "non-determinacy" and the community of physics expected that "God plays dice." Because of this historical reason, Newton was seen as a deterministic fatalist, Einstein and some others became old diehards against randomness and chance. The special goal was to employ "indeterminacy" to enter a "probabilistic universe," and to completely destroy the "Newtonian realm" by making use of the chaos theory. However, classical mechanics is in fact an invariant system of the type of Plato's static figures [Lin & OuYang, 2010; OuYang, McNeil, & Lin, 2002; Bergson, 1963; Koyré, 1968; Prigogine, 1980] without involving the concept of "determinacy". Considering the fact that in the universe the only thing that stays invariantly constant is change, neither the corresponding time series' stability of the stochastic system nor the well-posedness of the classical mechanical system can truly handle physical "indeterminacy" or "irregularity." Evidently, corresponding to the "invariant" system are variable events. Wishing for "indeterminacy" without knowing what "indeterminacy" is and where it is from can only make the concept of "indeterminacy" an imaginary fairy without its space for existence [OuYang & Lin, 2006].

When people are either vexed by or interested in chaos, they do not recognize that it is non-chaotic similarities that stand for the true chaos . That is what is truly and practically scary. If all people in the world were identical, then no one could find himself out of the ocean of the same-looking bodies. Fortunately, the world in which mankind exists allows differences so that me, you, him, ... can be easily distinguished. It seems that there is no such a need for us to be tied up on the indeterminacy of our conscious; instead we should focus more on the understanding of what indeterminate events are.

There is no denial that classical mechanics and the stochastic system of the mathematical logic cannot practically provide us with any badly needed "risk value," even though all we can do in front of risks is make use of the empirical statistics or experiments. That explains why in front of (nearly-) zero probability disasters, we still do not know what to do.

Similarly, there is no denial that the quantitative system, with which we have become so familiar, faces the challenge that variable events do not really comply with the rules of the calculus of formal logic of quantities. Thus it seems that what is more important for our time than creating "chaos" out of "deterministic equations" is how to understand and how to handle variable events.

To conclude this presentation, let us provide a final note. The purpose of the work presented in this paper is not to destroy Lorenz's chaos. In fact, we simply use Lorenz's model and his relevant work as a convenience to show that the current system of quantities and the system of modern science suffer from major difficulties in both theoretical analysis and practical applications. For details see [Lin, 1998; Wu & Lin, 2002; Lin & OuYang, 2010]. As with the case that although $1 + 1 = 2$ is not universally true, it does have a wide range of practical applications. And, although the system of modern mathematics in general and calculus in particular suffers from inconsistencies which have been seen as the fourth crisis in the foundations of mathematics [Lin, 2008], they have indeed helped to bring magnificent scientific advances to mankind. Therefore, it is not surprising to find successful applications of Lorenz's chaos theory in various specific situations (relevant reference are too numerous to list), because it does reveal the fact that when one employs quantitative means to study nature, one must face "chaos," because the mathematical tool used is too primitive to capture the essence of the nature one seeks to investigate. (This end has made chaos theory extremely welcomed in the communities of various religions). Whether or not the Lorenz's model is too simple to be a good representation of atmospheric convection is not a concern in our works, because no matter how complicated or sophisticated is the model, as long as it involves the symbols of differentiation, the model is only potentially valid on the basis of continuity and differentiability, which are two conditions rarely satisfied in nature. See [Lin, 2008b] for details. To this end, Lorenz [1996] described a system of over five million equations in over five million variables in a European effort to describe climate. Still, as expected, this system exhibits "chaos." As for whether the phenomenon of chaos is an artifact or not, the answer is both YES and NO. As shown in [Lin, 2008], nature contains such phenomena that quantities are just not powerful enough to describe. Additionally, we have addressed this question in [Lin, 1998] from such diverse angles as mathematics, physics, mathematical modeling, philosophy, evolution science, numerical experiments, and practical applications. That piece of work attracted well over one thousand communications from scholars all over the world during that first year of publication. Thus, to this end we claim that we and the colleagues in our research group are the first to have openly and comprehensively settled this last concern of whether or not chaos is a numerical artifact. For the relative completeness of our presentation, we mention specifically that OuYang and Wu [1998] demonstrated by employing mathematical group theory that just as Lorenz's model is deterministic, so are the solutions. More specifically, they showed that the three branching solutions of Lorenz's model can be connected together through the transformation group $\{I, e\}$, and that if an orbit of the model is given, there must be another corresponding orbit through the group $\{I, e\}$. Along similar lines using rigorous computer estimates, Tucker [1999] proved that the Lorenz attractor is indeed a chaotic attractor, as a numerical artifact.

## 1.5. Appendix: Another Way to Show that Chaos Theory Suffers From Flaws

For understanding the quantitative instability and for acquiring the intuition of the mathematical properties of nonlinear equations, let us look at the well-known nonlinear mathematical model of the Newtonian system:

$$\frac{du}{dt} = F = ku^2 \tag{A1}$$

Here we will analyze the essence of stability of this model using different initial values. Let us rewrite Eq. (A1) as a difference equation as follows:

$$u^{(n+1)} = u^{(n)} + k(u^{(n)})^2 \Delta t \tag{A2}$$

where the superscript $n$ stands for the time step. The stability of this problem can be discussed in three cases as follows.

Case 1: Considering the practical use, let us take the effective digits of the initial values to one digit after the decimal point. If we take

$$u^{(0)} = 10.1\ m/s;\ k = 1\ m^{-1}s^{-1};\ \Delta t = 10\ s$$

where $m$ stands for meter and $s$ second. Substituting these values into Eq. (A2) for $n = 1$ gives

$$u^{(1)} > 1000\ m/s$$

When $n = 2$, we have

$$u^{(2)} > 10^7\ m/s$$

$$\cdots\cdots\cdots\cdots\cdots$$

What is shown indicates that under the condition of non-small effective quantities, with only two steps of computation, the calculated result of nonlinear Eq. (A2) is already far greater than the first cosmic speed. That is, the moving article of the Newtonian dynamic system has already escaped the gravitation of the Earth flying into the outer space. That is the spirit of the quantitative computational instability of nonlinearity.

Case 2: Let us fully understand the quantitative computation problem of nonlinearity. In practical applications, there must be such inevitable situations where the initial values or parameters take on small effective values. To analyze this situation, let us once again look at some particular calculations by using the first digit after the decimal point as the effective place value. If we take

$$u^{(0)} = -0.1m/s,\ k = 1m^{-1}s^{-1}, \Delta t = 6s$$

Substituting these values into Eq. (A2) in turn produces

$$\begin{aligned} u^{(1)} &= -0.04\ m/s \\ u^{(2)} &= -0.0304\ m/s \\ u^{(3)} &= -0.0250\ m/s \end{aligned}$$

$$\cdots\cdots\cdots\cdots\cdots$$

Since we have predetermined the effective value to be the first digit place after the decimal point, all the previous results have fallen into the errors of the initial values. This is because digits beyond the first place after the decimal point no longer exist practically or can be ignored. Because of this reason, it is evident that quantities can be calculated that have meaningless accuracy. This example is a materialistic realization of the statement that there are things numbers cannot describe.

Each well-known numerical computation involving infinitesimal differences of large quantities contains the problem of computational inaccuracy of the effective values. However, one has to be clear that such situations may not be seen as the characteristic only nonlinearity possesses, because for linear or general equations involving differences of quasi-equal quantities such situations also exist. For instance, in Eq. (A2), if $u^{(0)} > 0$ and $k < 0$, even with non-small effective values, the computational results could also produce errors.

Case 3: If we rewrite Eq. (A1) as follows:

$$\frac{du}{dt} = ku^2 + c \tag{A3}$$

then its difference equation is

$$u^{(n+1)} = u^{(n)} + k(u^{(n)})^2\Delta t + c\Delta t$$

Let us take $u^{(0)} = 5\ m/s$; $k = 0.1\ m^{-1}s{-}1$; $c = -2.5; \Delta t = 0.1\ s$. Substituting these values into Eq. (A4) produces

$$\begin{aligned} u^{(1)} &= 5.0000\ m/s \\ u^{(2)} &= 5.0000\ m/s \\ u^{(3)} &= 5.0000\ m/s \\ &\cdots\cdots\cdots \end{aligned}$$

What is interesting is that since $c$ is a negative number, Eq. (A4) involves an infinitesimal difference of large quantities, leading to the result of attractors, as in chaos theory. From this example, the reader should be able to see the essence of chaos theory. That is, non-first-order products of small effective values or small difference of large quantities can suppress the nonlinear explosive growth. If in Case 2 one gets involved with irrational or repeating decimal numbers, then there will be further problems with computational accuracy. That is such a typical situation that there are matters numbers cannot describe.

## References


1. Bergson, H. [1963] "L´evolution créatrice," In: Oeuvres, Edtions du Centenaire, PUF, Paris, France.
2. Koyré, I. [1968] *Etudes Newtoniennes*, Paris: Gauimard.
3. Lin, Y. (ed.) [1998] "Mystery of Nonlinearity and Lorenz's Chaos. A special double issue of Kybernetes," *Inter. J. Cybernetics & Systems* **27**, 605–854.
4. Lin, Y. (ed.) [2008a] "Systematic studies: The infinity problem in modern mathematics. Kybernetes," *Inter. J. Cybernetics & Systems* **37**(3-4), 387–578.
5. Lin, Y. [2008b] *Systemic yoyos: Some Impacts of the Second Dimension*, Auerbach Publications, USA, an imprint of Taylor and Francis.
6. Lin, Y. & OuYang, S. C. [2010] *Irregularities and Prediction of Major Disasters*, Auerbuch Publications, USA, an imprint of Taylor and francis.
7. Lin, Y., OuYang, S. C., Li, M. J., Zhang, H. W. & Jiang, L. J. [2001] "On fundamental problems of the 'chaos' doctrine," *Inter. J. Appl. Mathematics* **5**, 37–64.
8. Lorenz, E. N. [1963] "Deterministic Nonperiodic Flow," *J. Atmos, Sci* **20**, 130–141.
9. Lorenz, E. N. [1996] *Essence of Chaos*, Washington University Press, USA.
10. OuYang, S. C. [1998] *Weather Evolutions and Prediction*, Meteorological Press, China.
11. OuYang, S. C. & Wu, Y. [1998] "The transformation group of branching solutions of Lorenz's equations and 'chaos'," *Inter. J. Cybernetics & Systems* **27**(6/7), 669–673.
12. OuYang, S. C., Chen, Y. G. & Lin, Y. [2009] *Information Digitization and Prediction*, Meteorological Press, China.
13. OuYang, S. C. & Lin, Y. [1997] "Some problems on population models and "one-dimensional iterative formula," *Systems Analysis Modelling Simulation* **28**, 223–233.
14. OuYang, S. C. & Lin, Y. [2006] "Disillusion of randomness and end of quantitative comparability," *Scientific Inquiry* **7**, 171–180.
15. OuYang, S. C., McNeil, D. H. & Lin, Y. [2002] *Entering the Era of Irregularity*, Meteorological Press, China.
16. Prigogine, I. [1980] *From Being to Becoming: Time and Complexity in the Physical Science*, W H Freeman, USA.
17. Saltzmann, B. [1962] "Finite amplitude free convection as an initial value problem," *Int. J. Atmos. Sci* **19**, 329–341.
18. Tucker, W. [1999] "The Lorenz attractor exists," *Comptes Rendus de l'Academie des Sciences Series I. Mathematique* **328**(12), 1197–1202.
19. Wu, Y. & Lin, Y. [2002] *Beyond Nonstructural Quantitative Analysis: Blown-Ups, Spinning Currents and the Modern Science*, World Scientific, USA.

# Chapter 2

# Nonexistence of Chaotic Solutions of Nonlinear Differential Equations

Lun-Shin Yao

*School of Mechanical, Aerospace, Chemical and Materials Engineering*
*Arizona State University Tempe, AZ 85287, USA*
*E-mail: ls_yao@asu.edu*

We discuss some important issues arising from computational efforts in dynamical systems and fluid dynamics. Various individuals have misunderstood these issues since the onset of these problem areas; indeed, they have been routinely misinterpreted, and even viewed as "laws" by some. This paper hopes to stimulate appropriate corrections and to realign thinking, with the overall goal being sound future progress in dynamical systems and fluid dynamics.

**Keywords**: Chaos, turbulence, instability, computational errors, differential equations.

## Contents



## 2.1. Introduction

Both existence and uniqueness require consideration when studying a system of differential equations. The existence of chaotic solutions for differential equations (not discrete algebraic maps) has never been proven; instead, existence is routinely taken for granted. However, differential equations presumed to be chaotic are unstable; therefore, no convergent numerical solutions for them are possible, as clearly pointed out by Von Neumann. It is likely that all published numerical solutions are simply numerical errors [Yao, 2007; Yao, 2010; Yao & Hughes, 2008a, 2008b]. The upshot of this conclusion is that there exists the frightening possibility that no solutions of chaotic differential equations are obtainable through any existing discretized numerical method, implying that solutions do not exist. Yao [2010] discussed this matter in detail for the Lorenz system. Without a resolution of the existence question, it seems frivolous to discuss properties of possibly non-existing solutions of differential equations. In fact, the relevance of Smale's horseshoe to differential equations is questionable. We discuss these matters in the first two open problems defined below.

The relevance of chaos to turbulence is an unsettled question. Chaos is random in time; turbulence is random in time and space. We believe they share some fundamental properties. On the other hand, multiple solutions of the Navier-Stokes equations exist when the Reynolds number is larger than its critical value; they reflect sensitivity to initial conditions [Yao, 2009]. The selection principles for an appropriate solution and the resulting nonlinear wave interactions for different initial conditions are explained by the resonant wave theory developed by Yao [1999]. However, multiple solutions of the Navier-Stokes equations evolve in an entirely different manner than the description of Smale's horseshoe; this leads to the third open problem discussed below. It also provides evidence that sensitivity to initial conditions cannot be a sufficient condition for chaos, as many practitioners of dynamical systems believe.

## 2.2. Open Problems About Nonexistence of Chaotic Solutions

**Problem 2.1.** *No chaotic solution of a system of nonlinear differential equations exists?*

The Poincaré-Bendixson Theorem clearly implies that a two-dimensional continuous dynamical system cannot give rise to a strange attractor. Some published chaotic solutions for a single nonlinear or linear ordinary differential equation violate this theorem. They are unstable spurious numerical solutions, but are frequently misinterpreted as chaos [Griffiths, Sweby & Yee, 1992]. Since no analytical chaotic solution is available for a system of more than two nonlinear differential equations, a numerical solution becomes the only alternative. Systems of ordinary differential equations that exhibit chaotic responses have yet to be correctly integrated. So far no convergent computational results have been confirmed for chaotic differential equations. Various computed numbers are not solutions of the continuous differential equations; all proposed chaotic responses are, more likely than not, simply numerical noise and have nothing to do with the solutions of differential equations. Two well-known Lorenz models [Lorenz, 1963, 1990] are typical examples showing that computed non-periodic solutions are simply due to the unstable amplification of truncation errors and the distribution of singularities [Teixeira, Reynolds & Judd, 2007; Yao & Hughes, 2008b; Yao, 2010].

Numerical methods convert continuous differential equations to a set of algebraic equations to be solved by computers. Derivatives in the continuous equations are replaced by corresponding discretized forms. For example $\frac{dx}{dt} \simeq \frac{\Delta x}{\Delta t}$ where $\frac{\Delta x}{\Delta t}$ equals $\frac{dx}{dt}$ in the limit as $\Delta t$ approaches zero. Instead of taking the limit, $\Delta t$ has a finite "small" value in the discretizing procedure such that $\frac{\Delta x}{\Delta t}$ sufficiently approximates $\frac{dx}{dt}$ when $\Delta t$ is "small." Von Neumann established that discretized algebraic equations must be consistent with the differential equations, and must be stable in order to obtain convergent numerical solutions for the given differential equations. This can be easily checked by ensuring that the difference between computed results for successively reduced time-step size is acceptably small. A successful check can be taken as an indication that $\frac{\Delta x}{\Delta t}$ is indeed a good approximation to $\frac{dx}{dt}$ and that the solutions of the discrete approximations have converged to the solution of the continuous equations. Any computed results that fail this check are simply numerical errors that have no mathematical meaning. This is the most important fundamental rule in solving differential equations numerically by computers. Unfortunately, this rule is consistently ignored in computational chaos and turbulence, and researchers in numerical chaos have often argued that it is irrelevant. One wonders how a set of discretized algebraic equations can be related to differential equations if convergence cannot be assured.

Yao and Hughes [2008b] repeated the Lorenz computation [2006a] and found that his numerical solutions do not converge. The positive Lyapunov exponents found in his computation are simply due to unstable computations, which violate Von Neumann's convergence criteria. It is worthwhile to note that Lorenz's second model [Lorenz, 1990] is not uniformly dissipative and does not have an attractor, but its trajectory settles inside a large attracting set. The unstable computation of Lorenz's first model [Lorenz, 1963] has been analyzed and discussed in detail by Yao [2010]. Some important misconceptions will be discussed below in Open Problem 2.2. Additional examples of unstable computation of nonlinear differential equations have also been discussed and can be found in [Yao, 2010]. Comprehensive and carefully computed examples that show useless computational results, which are integration time-step dependent, can be found in [Teixeira, Reynolds & Judd, 2007; Yao & Hughes, 2008a].

It would be an exciting contribution if a convergent computed chaotic solution for any system of nonlinear differential equations could be obtained. According to our experience, this would require a breakthrough in computational methods. A much larger and faster computer alone will not solve this problem. Beyond that, there is no reason to believe that a chaotic or turbulent solution exists for differential equations, including the Navier-Stokes equations in fluid mechanics.

**Problem 2.2.** *Is Smale's horseshoe relevant to chaotic solution of differential equations, including direct simulation of turbulence?*

Smale's horseshoe lays the foundation of the structural stability of dynamic systems, chaos. The studies in the sixties concluded that a hyperbolic dynamic system is structurally stable. In a hyperbolic system, the tangent space at every point can be decomposed into three complementary directions, which are stable, unstable and neutral. However, no system of differential equations is hyperbolic [Viana, 2000].

In the $14^{th}$ of 18 challenging mathematical problems [Smale, 1998], Smale asked: Is the dynamics of the ordinary differential equations of Lorenz [1963] that of the geometric Lorenz attractor of Williams, Guckenheimer and Yorke? One can view this as a general question about the equivalence of discrete algebraic maps and differential equations. In spite of numerous attempts, a convincing proof of the existence of the geometric Lorenz attractor for the Lorenz's differential equations was not achieved until Tucker [1999] provided a solution to this problem with the aid of his computer. He showed that:

(1) The Lorenz equations are dissipative; consequently, they contain a forward invariant region. This shows it is a closed invariant set.
(2) The result is sensitive to initial conditions so it is a strange attractor.
(3) The trajectory is indecomposable and looks topologically transitive.

On the other hand, Yao [2010] reported that computed chaotic solutions of the Lorenz equations are consistently contaminated by the amplification of truncation errors, and are sensitive to artificial integration time-steps. He carried out a thorough analysis of the first Lorenz model and showed that two mechanisms can amplify truncation errors. One error is well known; it is amplified along the unstable manifold and causes an exponential amplification of truncation errors. This violates the stability requirement of Von Neumann's convergence so the computational results have no value. However, if one follows Tucker's method, it can be shown that the numerical results, which are dependent on the integration time-steps, satisfy all three properties of Lorenz algebraic maps!

The second mechanism is the explosion amplification of truncation errors. This is because the trajectory penetrates the virtual separatrix, and violates the differential equations. The existence of a virtual separatrix is a consequence of singular points of a non-hyperbolic system of differential equations, which is not shadowable.

Sensitivity to initial conditions has been studied for hyperbolic systems [Bowen, 1975], and is an active topic for nearly hyperbolic systems [Dawson, Grebogi, Sauer, & Yorke, 1994; Viana, 2000; Palis, 2000; Morales, Pacifico & Pujals, 2002]. In particular, the methods of shadowing for hyperbolic systems have shown that trajectories may be locally sensitive to initial conditions while being globally insensitive since true trajectories with adjusted initial conditions exist. Such trajectories are called shadowing trajectories, and lie very close to the long-time computed trajectories. However, systems of differential equations arising from physical applications are not hyperbolic systems. If the attractor is transitive (ergodic), all trajectories are inside the attractor so they are generally believed to be solutions of the underlying differential equations, whatever the initial conditions. Little is known about systems, which are not transitive (non-ergodic). Do & Lai [2004] have provided a comprehensive review of previous work, and discussed the fundamental dynamical process indicating that a long-time shadowing of non-hyperbolic systems is not possible.

Another commonly cited computational example in chaos involves two solutions of slightly different initial conditions that remain "close" for some time interval and then diverge abruptly. In fact, this behavior is often believed to be a typical characteristic of chaos. More properly, this phenomenon is actually due to the explosive amplification of numerical errors, and violation of the differential equations as described above. Similar computational results can occur for two different integration time steps; many would consider such results as acceptable since it is a "twin brother" of sensitivity to initial conditions and a typical characteristic of chaos. This mistake deserves clarification.

In fact, two mechanisms of error amplification can alter the topological properties of attractors. This shows the common expectation that the presence of an attractor guarantees the correctness of computed results is completely wrong. It is easy to conclude that Tucker did not prove that the Lorenz system has the property of a Lorenz map since a solution of the Lorenz system does not exist.

Before we move to the next open problem, we would like to point out that the inviscid-flow equations do not have non-periodic turbulent solutions. This is because these equations are not dissipative. Those studying chaos in conservative systems, which cannot have attractors, should check their computations to ensure that their results are not dependent on the integration time-step.

**Problem 2.3.** *What is the relation between Smale's horseshoe and the physics of the Navier-Stokes equations?*

That the structure of resonances determines the long-time solutions of nonlinear differential equations was originally advocated by Stokes [1847] and Poincaré [1892]; the later developed the well-known Poincaré-Lighthill-Kuo (PLK) Method. Yao [1999] proposed an eigenfunction-spectral method, a fully nonlinear approach and a generalization of the PLK method, for a direct numerical solution of the Navier-Stokes equations. In the following discussion, the generalized eigenfunctions are referred to as "waves". The linear terms of the Navier-Stokes equations involve the local acceleration terms, the viscous dissipative terms, and the pressure-gradient terms. The nonlinear terms represent the energy transfer among all waves, traditionally referred to as inertial effects. There are two kinds of transfer: forced transfer and resonance transfer. The forced transfer can create a new wave, but with limited energy transfer. Substantial energy transfer can only occur when the resonance conditions are satisfied [Phillips, 1960].

The basic idea of the eigenfunction-spectral method is that a solution of nonlinear differential equations can be expressed in terms of a complete set of waves. The amplitude of every wave is not necessarily nonzero. The base flow is a special wave, which has zero frequency and infinite wavelength. The pressure gradient drives the base flow, and is a source of energy. For the Reynolds number, *Re*, less than its critical value, all disturbance waves are stable and highly dissipative; thus, only the base flow exists. As Re reaches its critical value, the energy received by the critical wave from the base flow balances with dissipation so its amplitude does not change with respect to time. This is the condition of neutral stability, or the onset of instability. Any further increase of *Re*, the energy from the base flow becomes larger than that can be dissipated by waves; the amplitude of the waves will grow. As *Re* increases further, more waves become unstable and the nonlinear wave interaction becomes complicated. This is usually called finite-amplitude instability. We found that convergent computation becomes more difficult, and computations eventually diverge when $Re$ increases further, but is still far below its transitional value for the flow to become turbulent. This fact is not well known and respected; many researchers routinely compute finite-amplitude instabilities and turbulence, generating numerical errors, without bothering to check their convergence.

The nonlinear interactions among Taylor-Couette vortices (driven by boundary motion) with different wave numbers were studied by using a weakly nonlinear theory of multiple waves [Yao & Ghosh Moulic, 1995a] and nonlinear theory [Yao & Ghosh Moulic, 1995b]. Both theories can be derived from the eigenfunction-spectral method as truncated solutions. They represent the disturbance by a Fourier integral and derive an integro-differential equation for the evolution of the amplitude density function of a continuous spectrum. Their formulation allows nonlinear energy transfer among all participating waves and remedies the shortcomings of equations weakly nonlinear theories of a single wave of the Landau-Stuart type. Numerical integrations of this integro-differential equation indicate that the equilibrium state of the flow depends on the wave number and amplitude of the initial disturbance as observed experimentally and cannot be determined uniquely on each stable bifurcation branch without knowing its history. The accessible wave numbers of the nonlinear stable range lie within the linear unstable range minus a small band, which was traditionally considered as the range of side-band instability. The numerical results simply indicate that waves in this small band lose more energy than they gain; therefore, waves within this small band decay and excite their harmonics. This agrees with Snyder's observation [Snyder, 1969] that the trend of energy transfer is determined by the stability characteristics of the base flow. It is a consequence of fully nonlinear wave interactions, and is not due to sideband instability as speculated by weakly nonlinear theories of a single wave; this misconception has prevailed for a long time, and is still pursued by some classical fluid dynamists.

The analysis of mixed convection in a vertical annulus using a nonlinear theory with continuous spectrum [Yao & Ghosh Moulic, 1994, 1995b] yields a conclusion identical to that from the study of Taylor-Couette flows. They treated the spatial problem as a temporal problem. This is possible by following a control-mass system. Plots of the evolution of the kinetic-energy spectra for the two problems are identical in shape and differ only in amplitudes and ranges of wave numbers. This indicates that the evolution of energy spectra provides universal and fundamental information for all flows. The selection of the equilibrium wave number is a result of nonlinear wave interactions.

The principles of nonlinear interaction of participating waves are universal and fundamental, and are shared by all flows. The mechanism that allows multiple solutions to exist and the required associated conditions have been demonstrated. The selection principles, due to its property of sensitive to initial conditions, deduced from the numerical study of [Ghosh Moulic & Yao, 1996; Yao, 1999; Yao & Ghosh Moulic, 1995a; 1995b], are listed below:

(1) When the initial disturbance consists of a single dominant wave within the nonlinear stable region, the initial wave remains dominant in the final equilibrium state. Consequently, for a slowly starting flow, the critical wave is likely to be dominant.
(2) When the initial condition consists of two waves with finite amplitudes in the nonlinear stable region, the final dominant wave is the one with the larger initial amplitude. If the two waves have the same initial finite amplitude, the dominant wave will be the one closer to the critical wave. On the other hand, if the initial amplitudes are very small, the faster growing wave becomes dominant.
(3) When the initial disturbance has a uniform broadband spectrum of noise, the final dominant wave is the fastest linearly growing wave, if the initial amplitude is small. On the other hand, if the uniform noise level is not small, the critical wave is the dominant equilibrium wave.
(4) Any initial disturbance outside the accessible frequency range will excite its sub-harmonics or super-harmonics, dependent on the existence of broadband noise, whichever are inside the accessible frequency range. The accessible frequency range is the linear stability range minus a small band.

Similar principles of nonlinear energy transfer have also been found in other simple nonlinear partial differential equations [Yao, 2007, 2009]; thus, the discussion outlined above is general for all nonlinear differential equations. The open question is: Does a one-to-one correspondence exist between resonance and Smale's horseshoe?

From the above discussion, one can easily conclude that a sensitivity-to-initial condition is not sufficient for the existence of a strange attractor since such conditions are also true for all instabilities when the *Re* is larger than its critical value. A strange attractor may exist only if the dynamics have a closed invariant set, and it is sensitive to initial conditions. On the other hand, Open Problem 1 demonstrates that the existence of a strange attractor does not guarantee the correctness of the computed solution. The fact is that many numerical generated attractors are simply the consequence of amplification of truncation errors associated with discrete numerical methods.

Another interesting, but not well known example showing the importance of nonlinearity of the Navier-Stokes equations is related to the calculation of the critical Reynolds number. The critical *Re* of linear-stability theory (the linear-stability limit) is the value of *Re* above which ever-present single infinitesimal critical wave grows. Equivalently, a critical *Re* can be defined when the initial amplitude of the critical wave does not change with time as described above. The linear-stability theory is a single-wave theory. The physics of flow instability agrees with the nonlinear theory involving many waves, and differs from the linear theory. The critical *Re*, calculated by the nonlinear theory, could be a function of the initial amplitude of the critical wave, since energy can be transferred to harmonics and sub-harmonics of the critical wave. Without Considering nonlinear effects, all disturbances at the critical *Re*, calculated by a linear-stability analysis, will decay to zero [Yao, 2007].

It is worth noting that conventional studies of fluid mechanics often relied on flow visualization and simple analysis, which can only provide limited information about a particular base flow. In particular, the details of the eigenfunctions discussed earlier, which are important to an understanding of the stability characteristics of a base flow, cannot be total revealed by relatively crude visualization methods and simple analysis. Thus, in the study of certain flows, early researchers may have concluded their work without "seeing" all important features of the flow, causing them to overlook the existence of other possible solutions. The principle of multiple solutions is an important aspect of fluid mechanics that has been largely overlooked [Yao, 2009]. This character of multiple solutions is shared by all nonlinear differential equations as a consequence that their solutions are sensitive to initial conditions.

Acknowledgement: The author wish to thank his colleague Professor Steve Baer, Mathematics at ASU for his continuous interest and discussion of chaos for the past 15 years.

## References

1. Bowen, R. [1975] "Equilibrium states and the ergodic theory of Anosov diffeomorphism," *Lecture Notes in Mathematics* **470**, Spring-Verlag.
2. Dawson S., Grebogi C., Sauer T. & Yorke J. A. [1994] "Obstructions to shadowing when a Lyapunov exponent fluctuates about zero," *Phys. Rev. Letters* **73**, 1927–1930.
3. Do, Y. & Lai, Y. C. [2004] "Statistics of shadowing time in non-hyperbolic chaotic systems with unstable dimension variability," *Phys. Rev. E* **69**, 016213.
4. Ghosh Moulic, S. & Yao, L. S. [1996] "Taylor-Couette instability of traveling waves with a continuous spectrum," *J. Fluid Mech* **324**, 181–198.
5. Griffiths, D. F., Sweby, P. K. & Yee, H. C [1992] "On spurious asymptotic numerical solutions of explicit Runge-Kutta methods," *IMA. J. Numerical Analysis* **12**(3), 319–338.
6. Lorenz, E. N. [1963] "Deterministic Nonperiodic Flow," *J. Atmos. Sci* **20** 130–141.
7. Lorenz, E. N. [1990] "Can Chaos and Intransitivity lead to Interannual Variability?," *Tellus* **42A**, 378–389.
8. Lorenz, E. N. [2006a] "Computational periodicity as observed in a simple system," *Tellus* **58A**, 549–557.
9. Lorenz, E. N. [2006b] "An attractor embedded in the atmosphere," *Tellus* **58A**, 425–429.
10. Morales, C. A., Pacifico M. J., & Pujals E. R. [2004] "Robust transitive singular sets for 3-flows are partially hyperbolic attractors or repellers," *Annals of Mathematics* **160**, 375–432.
11. Palis J. [2000] "A global view of dynamics and a conjecture on the denseness of attractors," *Astérisque* **261**, 335–347.
12. Palis J. & de Melo W. [1982] *Geometric Theory of Dynamical Systems*, Springer-Verlag. New York.
13. Phillips, O. M. [1960] "On the dynamics of unsteady gravity waves of finite amplitude. Part I," *J. Fluid Mech* **9**, 193–217.
14. Poincaré, H. [1892] *Les Méthods Nouvelles de la Mécanique Célèste* **I**, Gauthier-Villars, Paris.
15. Smale, S. [1998] "Mathematical problems for the next century," *The Mathematical Intelligencer* **20**, 7–15.
16. Snyder, H. A. [1969] "Wave-number selection at finite amplitude in rotating Couette flow," *J. Fluid Mech* **35**, 273–298.
17. Stokes G. G. [1847] "On the Theory of Oscillatory Waves," *Cambridge, Transactions* **8**, 441–473.
18. Teixeira, J., C. A. Reynolds, & K. J. [2007] "Time step sensitivity of nonlinear atmospheric models: numerical convergence, truncation error growth, and ensemble design," *J. Atmos. Sci.* **64**, 175–189.
19. Tucker, W. [1999] "The Lorenz attractor exists," *Comptes Rendus de l'Academie des Sciences Series I. Mathematique* **328**(12), 1197–1202.

20. Viana, M. [2000] "What's new on Lorenz strange attractors?," *The Mathematical Intelligencer* **22**, 6–19.
21. Yao, L. S. [1999] "A resonant wave theory," *J. Fluid Mech* **395**, 237–251.
22. Yao, L. S. [2007] "Is a direct numerical simulation of chaos possible? A study of a model non-Linearity," *Inter. J. Heat & Mass Transfer* **50**, 2200–2207.
23. Yao, L. S. [2009] "Multiple solutions in fluid mechanics," *Non. Analysis: Model. & Control* **14**(2), 263–279. [http://www.lana.lt/journal/issues.php].
24. Yao, L. S. [2010] "Computed chaos or numerical errors," *Non. Analysis: Model. & Control* **15**(1), 109–126. [http://www.lana.lt/journal/issues.php].
25. Yao, L. S. & Ghosh, M. S. [1994] "Uncertainty of convection," *Inter. J. Heat & Mass Transfer* **37** 1713–1721.
26. Yao, L. S. & Ghosh, M. S. [1995a] "Taylor-Couette instability with a continuous spectrum," *J. Appl. Mech* **62**, 915–923.
27. Yao, L. S. & Ghosh, M. S. [1995b] "Nonlinear instability of travelling waves with a continuous spectrum," *Inter. J. Heat & Mass Transfer* **38**,1751–1772.
28. Yao, L. S. & Hughes, D. [2008a] "Comment on "Time step sensitivity of nonlinear atmospheric models: Numerical convergence, truncation error growth, and ensemble design" Teixeira *et al.*,(2007)," *J. Atmos. Sci* **65**(2), 681–682.
29. Yao, L. S. & Hughes, D. [2008b] "A comment on "Computational periodicity as observed in a simple system," by Edward N. Lorenz (2006a)," *Tellus* **60A**(4), 803–805.

# Chapter 3

## Some Open Problems in the Dynamics of Quadratic and Higher Degree Polynomial ODE Systems

F. Zhang[1] and Jack Heidel[2]

[1] *Department of Mathematics, Cheyney University of PA*
*E-mail: fzhang16@yahoo.com, fzhang@cheyney.edu*

[2] *Department of Mathematics, University of Nebraska at Omaha*
*Omaha NE, 68182*
*E-mail: jheidel@unomaha.edu*

Here we propose six open problems in dynamical systems and chaos theory. The first open problem is concerned with a rigorous proof of a collection of quadratic ODE systems being non-chaotic. The second problem is for a universal definition of non-chaotic solutions. The third problem is about the number of systems that can have chaotic solutions when the right hand sides are polynomials. The fourth problem is topologically how complicated a 2D invariant manifold has to be to contain and/or attract chaotic solutions. The fifth open problem is to show that a specific system has a solution with a fractal dimension on one of the Poincaré sections. The sixth problem is on rigorous proof of existence of chaotic solutions of some systems which exhibit chaos in numerical solutions.

**Keywords**: Rigorous proof of chaos, non-chaotic solutions, number of systems, invariant manifold, fractal dimension.

### Contents

### 3.1. First Open Problem

Several years ago Zhang and Heidel [Heidel & Zhang, 1999] and [Zhang & Heidel, 1997] showed that (almost) all dissipative and conservative three dimensional autonomous quadratic systems of ordinary differential equations with at most four terms on the right hand sides of the equations are non-chaotic. The sole exception is the system

$$\begin{cases} x' = y^2 - z^2 \\ y' = x \\ z' = y \end{cases} \tag{3.1}$$

which does however appear numerically to have a single unstable periodic solution and is therefore conjectured to also be non-chaotic. The above system is equivalent to the third order or "jerk" equation $z''' = z'^2 - z^2$. Very recently Malasoma [Melasoma, 2009] has shown that every jerk equation $z''' = j(z, z', z'')$ where $j$ is a quadratic polynomial with at most two terms, with the sole exception of (3.1), is non-chaotic. Thus carefully determining the behavior of solutions of equation (3.1) becomes an interesting problem instead of just being a passing curiosity. Recently Heidel and Zhang [Heidel & Zhang, 2007, Zhang & Heidel, 2010] showed that dissipative and conservative three dimensional autonomous quadratic systems of ordinary differential equations with five terms and one nonlinear term on the right hand sides of the equations have one system and four systems respectively that exhibit chaos. Most systems in [Heidel & Zhang, 2007] and [Zhang & Heidel, 2010] are proved to be non-chaotic. The remaining 20 systems which are listed here are proved to be non-chaotic when the parameter in each of the systems is in certain range. Extensive computer simulations indicate that there are no chaotic attractors in these systems. We conjecture that they are non-chaotic systems. Dissipative systems:

$$\begin{cases} x' = \pm x + y + Az \\ y' = xz \\ z' = y \end{cases} \tag{3.2}$$

$$\begin{cases} x' = y^2 \pm x \\ y' = x + Az \\ z' = y \end{cases} \tag{3.3}$$

$$\begin{cases} x' = y^2 + Az \\ y' = x \pm y \\ z' = x \end{cases} \tag{3.4}$$

$$\begin{cases} x' = y^2 + Az \\ y' = \pm y + z \\ z' = x \end{cases} \tag{3.5}$$

$$\begin{cases} x' = z^2 \pm x \\ y' = x + Az \\ z' = y \end{cases} \tag{3.6}$$

$$\begin{cases} x' = yz \pm x \\ y' = x + Az \\ z' = y \end{cases} \tag{3.7}$$

$$\begin{cases} x' = yz + Ay \\ y' = \pm y + z \\ z' = x \end{cases} \tag{3.8}$$

$$\begin{cases} x' = yz + Az \\ y' = x \pm y \\ z' = x \end{cases} \tag{3.9}$$

$$\begin{cases} x' = yz + Az \\ y' = \pm y + z \\ z' = x \end{cases} \tag{3.10}$$

$$\begin{cases} x' = \pm x + z \\ y' = Ay + z \\ z' = xy \end{cases} \tag{3.11}$$

$$\begin{cases} x' = \pm x + y + A \\ y' = xz \\ z' = y \end{cases} \tag{3.12}$$

$$\begin{cases} x' = \pm x + z + A \\ y' = xz \\ z' = y \end{cases} \tag{3.13}$$

$$\begin{cases} x' = yz + A \\ y' = x \pm y \\ z' = x \end{cases} \tag{3.14}$$

$$\begin{cases} x' = yz + A \\ y' = \pm y + z \\ z' = x \end{cases} \tag{3.15}$$

$$\begin{cases} x' = yz \pm x \\ y' = z + A \\ z' = x \end{cases} \tag{3.16}$$

$$\begin{cases} x' = \pm x + z \\ y' = x + A \\ z' = xy \end{cases} \tag{3.17}$$

$$\begin{cases} x' = \pm x + z \\ y' = z + A \\ z' = xy \end{cases} \tag{3.18}$$

conservative systems:

$$\begin{cases} x' = yz + Ay, \quad A < 0, \text{ for } +, A > 0, \text{ for } - \\ y' = \pm x + z \\ z' = x \end{cases} \tag{3.19}$$

$$\begin{cases} x' = y + z \\ y' = -x + Az \\ z' = xy \end{cases} \tag{3.20}$$

$$\begin{cases} x' = yz + A, \quad A < 0 \\ y' = x \pm z \\ z' = x \end{cases} \tag{3.21}$$

## 3.2. Second Open Problem

Ever since the chaotic attractor in the Lorenz equations [Lorenz, 1963] was discovered, chaos theory has become a popular branch in dynamical systems which attracts many mathematicians in the area. The central problems in chaos theory have been the definition, the mathematical properties, and analytic proof of existence of chaotic solutions, and discovery of possible chaotic systems and possible geometric patterns of chaos by numerical simulations. It is well known that among the central problems giving a universal definition of chaos in a mathematical sense for the solutions of dynamical systems is difficult because of the complexity of chaotic solutions. There are numerous definitions of chaos. Each definition emphasizes certain aspect(s) of a solution. Therefore none of them can be a general definition at this point. In [Brown & Chua, 1996] Brown and Chua listed nine definitions of chaos. For the nine definitions see the references therein. Here we give a tenth definition: A dynamical system is chaotic when it has an attractor with fractal dimension. Giving a universal definition of non-chaotic solutions is another way to define chaotic solutions. It is known that it is also very difficult. In [Heidel & Zhang, 2007] we gave a conjecture on the criterion recognizing non-chaotic behavior:
Consider the autonomous system

$$x' = f(x), \quad x \in \mathbb{R}^N, \quad t \in [0, \infty) \tag{3.22}$$

where $' = \frac{d}{dt}$, $f : \mathbb{R}^N \to \mathbb{R}^N$ is continuous. Let $x(0) = x_0$, and $x_j$, $x_{0j}$ and $f_j$, $j = 1, 2, ..., N$ be the components of $x$, $x_0$ and $f$ respectively.

**Criterion 3.1.** An $N$ dimensional system (3.22) with no cluster points in the set of isolated fixed points has no bounded chaos if for any of its solutions there are $N-2$ components $x_{n_k}(t), n_k \in 1, ..., N$ and $k = 1, ..., N-2$, such that for each of the $N-2$ components only the following cases can happen:

as $t \to \infty$ or $t \to -\infty$,

($i$) It tends to a finite limit.
($ii$) It is a periodic or asymptotic to a periodic function.
($iii$) It is unbounded.

there exists an $\omega$, $|\omega| < \infty$, such that,

($iv$) It is unbounded, $t \to \omega$,
($v$) It is bounded but does not have a limit, $t \to \omega$,
($vi$) It is bounded and has a limit, $t \to \omega$, but not defined at $t = \omega$.

When $N = 3$ this criterion has been widely accepted. For $N > 3$ it still needs verification. Even if this criterion is very useful, it misses countless types of non–chaotic solutions. This open problem is to give a universal definition of non–chaotic solutions.

### 3.3. Third Open Problem

It is well known that three-dimensional quadratic autonomous systems are the simplest type of ordinary differential equations in which it is possible to exhibit chaotic behavior. The Lorenz equations [Lorenz, 1963] and Rössler system [Rössler, 1976] both with seven terms on the right-hand side do exhibit chaos for certain parameter values. By computer simulation in [Sprott, 1994-1997, 2000a] J. C. Sprott found numerous cases of chaos in systems with five or six terms on the right-hand side. Heidel and Zhang showed in [Zhang & Heidel, 1997] and [Heidel & Zhang, 1999] that three-dimensional quadratic autonomous conservative and dissipative systems with four terms on the right hand side have no chaos. So a chaotic three-dimensional quadratic autonomous conservative or dissipative system must have at least five terms on the right hand side. In [Heidel & Zhang, 2007] the authors proved a general theorem in determining a non-chaotic solution and showed that among all three-dimensional quadratic autonomous conservative systems with five terms on the right hand side and one nonlinear term there is at most one of them that can have chaotic solutions. In [Zhang & Heidel, 2010] the authors show that there are only four examples of five-one dissipative systems that exhibit chaos. Let $P(x) = \sum_{|\alpha| \leqslant k} A_\alpha x^\alpha$ be a polynomial, where $x \in \mathbb{R}^N$ and $N \geqslant 1$ is an integer, $\alpha = (\alpha_1, ..., \alpha_N)$ and each of the $\alpha_i$ is a nonnegative integer, $x^\alpha = x_1^{\alpha_1} ... x_N^{\alpha_N}$, the order of the multi-index $\alpha$ is denoted by $|\alpha| = \alpha_1 + ... + \alpha_N$ and $A_\alpha \in \mathbb{R}$. Consider the autonomous system

$$x_i' = P_i(x) = \sum_{|\alpha^i| \leqslant k} A_{\alpha^i} x^{\alpha^i}, \quad i = 1, 2, ..., N \tag{3.23}$$

1. For a given $k$ and $N = 3$ after eliminating equivalent systems under the scalar transformations

$$x_1 = aX_1, \qquad x_2 = bX_2, \qquad x_3 = cX_3, \qquad t = \delta\tau$$

where $a$, $b$, $c$ and $\delta$ are nonzero real numbers and third-order permutation groups $P_g$, where $P_g$ has six elements

$$P_1 = \begin{pmatrix} 1\,0\,0 \\ 0\,1\,0 \\ 0\,0\,1 \end{pmatrix}, \quad P_2 = \begin{pmatrix} 0\,1\,0 \\ 1\,0\,0 \\ 0\,0\,1 \end{pmatrix}, \quad P_3 = \begin{pmatrix} 0\,0\,1 \\ 0\,1\,0 \\ 1\,0\,0 \end{pmatrix}$$

$$P_4 = \begin{pmatrix} 1\,0\,0 \\ 0\,0\,1 \\ 0\,1\,0 \end{pmatrix}, \quad P_5 = \begin{pmatrix} 0\,1\,0 \\ 0\,0\,1 \\ 1\,0\,0 \end{pmatrix}, \quad P_6 = \begin{pmatrix} 0\,0\,1 \\ 1\,0\,0 \\ 0\,1\,0 \end{pmatrix}$$

which systems in the form of (3.23) have chaotic solutions and how many systems are chaotic? Among those chaotic systems how many of them are conservative, and how many of them are dissipative?

2. For any given integer $N > 3$ and $k$ after eliminating equivalent systems under the scalar transformations

$$x_1 = a_1X_1, \quad x_2 = a_2X_2, \quad ..., \quad x_n = a_nX_n, \quad t = \delta\tau$$

and $n^{th}$-order permutation groups $P_g$, where $P_g$ has $m = n!$ elements

$$P_1 = \begin{pmatrix} 1\,0 \ldots & 0 \\ 0\,1 \ldots & 0 \\ \vdots\;\vdots\;\ddots & \vdots \\ 0\,0 \ldots & 1 \end{pmatrix}, \quad P_2 = \begin{pmatrix} 0\,1 \ldots & 0 \\ 0\,0 \ldots & 1 \\ \vdots\;\vdots\;\ddots & \vdots \\ 1\,0 \ldots & 0 \end{pmatrix}, \quad ..., \quad P_m = \begin{pmatrix} 0 \ldots & 0\,1 \\ 0 \ldots & 1\,0 \\ \vdots\;\ddots & \vdots\;\vdots \\ 1 \ldots & 0\,0 \end{pmatrix}$$

which systems in the form of (3.23) have chaotic solutions and how many systems are chaotic? Among those chaotic systems how many of them are conservative, and how many of them are dissipative?

3. More generally for a given $N \geq 3$ after eliminating equivalent systems under affine transformations

$$x' = Ay + b, \qquad t = \delta\tau, \qquad ' = \frac{d}{dt}$$

where

$$x = \begin{pmatrix} x_1 \\ x_2 \\ \vdots \\ x_n \end{pmatrix}, \quad y = \begin{pmatrix} y_1 \\ y_2 \\ \vdots \\ y_n \end{pmatrix}, \quad A = \begin{pmatrix} a_{11} & a_{12} & \dots & a_{1n} \\ a_{21} & a_{22} & \dots & a_{2n} \\ \vdots & \vdots & \ddots & \vdots \\ a_{n1} & a_{n2} & \dots & a_{nn} \end{pmatrix}, \quad b = \begin{pmatrix} b_1 \\ b_2 \\ \vdots \\ b_n \end{pmatrix}$$

which systems in the form of (3.23) have chaotic solutions and how many systems are chaotic? Among those chaotic systems how many of them are conservative, and how many of them are dissipative?

4. A related open problem is what types of chaotic attractors can the systems (3.23) have? Two examples of "types" of attractors are Lorenz attractor and Rössler attractor.

### 3.4. Fourth Open Problem

Consider $x' = f(x)$, $x \in \mathbb{R}^3$, where $f(x)$ are polynomials or $f \in C^n(\mathbb{R}^3)$, where $n$ is a nonnegative integer. If a solution of the system is asymptotic to a 2D $C^r$, $r \geq 1$ an integer, invariant manifold, can the solution be chaotic? If such chaotic solutions exists, topologically how complicated the 2D manifold has to be? What we mean by a manifold being complicated topologically is that for example a torus can be considered more complicated than a plane.

In particular, can a quadratic differential equation system on a torus exhibit chaos? In general can a solution of a system on a torus be more complicated than a space filling curve?

The following four figures are from [Heidel & Zhang, 2007]. Fig. 3.1 shows an orbit of system (3.24). It appears that the solution approaches a 2D surface which is topologically more complicated than a torus. From Fig. 3.2 and Fig. 3.4 even if they are not Poincaré sections of one solution, one can still tell from them that there are solutions in both cases that approach very complicated surfaces.

$$x' = y^2 - z + A, y' = z, z' = x \tag{3.24}$$

$$x' = y, y' = -x + yz, z' = 1 - y^2 \tag{3.25}$$

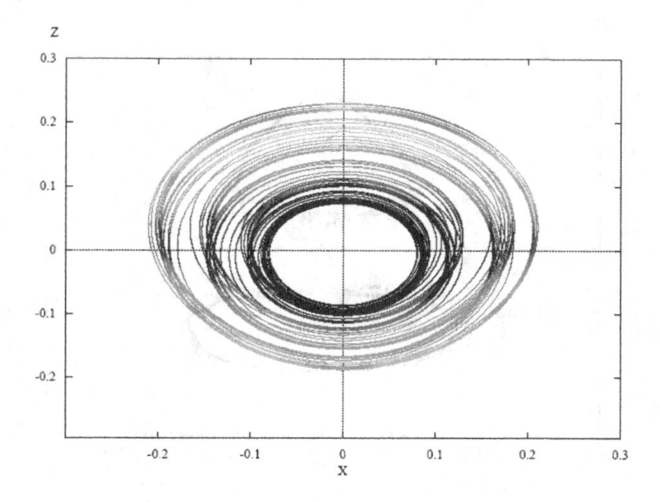


Fig. 3.1. (3.24)'s "chaotic" orbit, $x(0)$ =, 0.0428571, $y(0) = -0.105714$, $z(0) = -0.102325$, $A = -0.0125$.

## 3.5. Fifth Open Problem

From Fig. 3.2, and Fig. 3.3 it appears that system (3.24) has at least one solution such that it intersects the Poincaré section $x = 0$ into a set with "thickness". Similarly from Fig. 3.4, it appears that system (3.25) has at least one solution that intersects the Poincaré section $z = 0$ into a set also with "thickness". This open problem is to show that these sets have fractal dimensions. A further question is to show that this system is chaotic.

## 3.6. Sixth Open Problem

Chaotic solutions were proved to exist in some famous systems such as the Lorenz equations [Hassard *et al.*, 1994, Hastings & Troy, 1996] and Chua's circuit with piece-wise nonlinearity [Chua *et al.*, 1986]. It is well known that a rigorous proof of existence of chaotic solutions in a dynamical system is generally very difficult. In a Chua's circuit with smooth nonlinearity, there appear not only butterfly-like chaotic attractors, but also "small" chaotic attractors around equilibria for certain parameter regimes. It appears that the "small" chaotic attractor is of Rössler-type. No one has proved the existence of such small chaotic attractors. Sprott discovered numerous ODE systems having chaos [Sprott, 1997, 2000b]. Among them the following

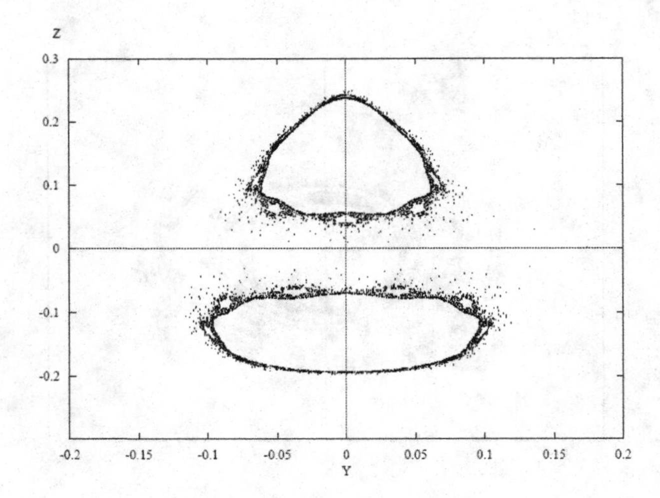


Fig. 3.2. (3.24)'s Poincaré section at $x = 0$ with initial condition $x(0) = 0.01$, $y(0) = -0.105714$, $z(0) = -0.102325$, $A = -0.0125$.

four systems are later studied in our work [Zhang & Heidel, 2010] on three dimensional dissipative autonomous quadratic systems with five terms and one nonlinear term on the right hand sides of the equations. System (3.27) and system (3.28) share the same scalar equation $u''' = uu' - u'' - Au$ in $y$ and $z$ respectively. The attractors are all of Rössler type. The open problem is to prove rigorously the existence of chaotic solutions in these systems:

$$\begin{cases} x' = y^2 - x + Az \\ y' = x \\ z' = y \end{cases} \tag{3.26}$$

$$\begin{cases} x' = yz - x + Ay \\ y' = z \\ z' = x \end{cases} \tag{3.27}$$

$$\begin{cases} x' = yz + Az \\ y' = x - y \\ z' = y \end{cases} \tag{3.28}$$

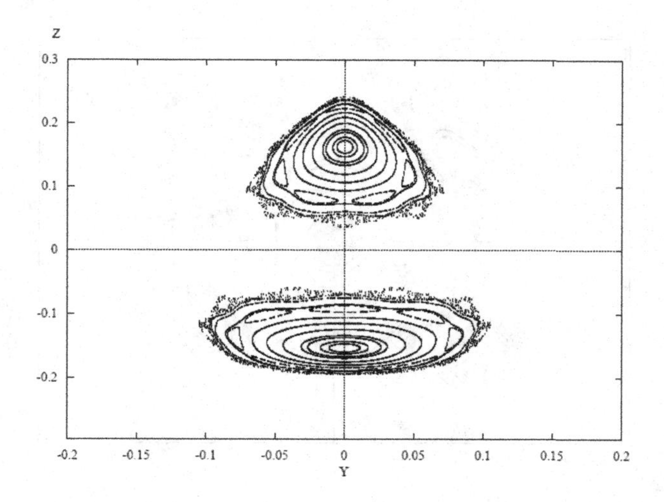


Fig. 3.3. (3.24)'s Poincaré section at $x = 0$ with initial condition $x(0) = 0.01$, $y(0) = -0.105714$, $z(0) = -0.102325$, $A = -0.0125$.

$$\begin{cases} x' = y^2 + Az \\ y' = x - y \\ z' = y \end{cases} \tag{3.29}$$

Other systems that Sprott discovered chaotic solutions include the following systems. The parameter values when the system exhibit chaos are given in [Sprott, 1994]. Again we are seeking for rigorous proof of the existence of chaotic solutions.

$$\begin{cases} x' = yz \\ y' = x - Ay, \quad A = 1 \\ z' = 1 - xy \end{cases} \tag{3.30}$$

$$\begin{cases} x' = yz \\ y' = x - Ay, \quad A = 1 \\ z' = 1 - x^2 \end{cases} \tag{3.31}$$

$$\begin{cases} x' = -y \\ y' = x + z, \qquad A = 3 \\ z' = xz + Ay^2 \end{cases} \tag{3.32}$$

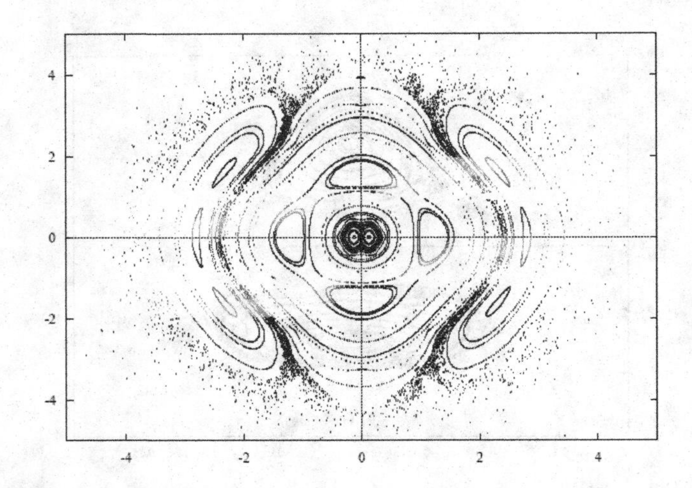


Fig. 3.4. Poincaré section of (3.25) at $z = 0$ with 37 different initial conditions $-2.5 \leqslant x(0) \leqslant 2.5$, $1.25 \leqslant y(0) \leqslant 5.25$, $z(0) = 0$.

$$\begin{cases} x' = yz \\ y' = x^2 - y, \quad A = 4 \\ z' = 1 - Ax \end{cases} \tag{3.33}$$

$$\begin{cases} x' = y + z \\ y' = -x + Ay, \quad A = 0.5, B = 1 \\ z' = x^2 - Bz \end{cases} \tag{3.34}$$

$$\begin{cases} x' = Ax + z \\ y' = xz - By, \quad A = 0.4, B = 1 \\ z' = -x + y \end{cases} \tag{3.35}$$

$$\begin{cases} x' = -y + z^2 \\ y' = x + Ay, \quad A = 0.5, B = 1 \\ z' = x - Bz \end{cases} \tag{3.36}$$

$$\begin{cases} x' = Ay \\ y' = x + z, \quad A = -0.2, B = 1 \\ z' = x + y^2 - Bz \end{cases} \tag{3.37}$$

$$\begin{cases} x' = Az \\ y' = By + z, \quad A = 2, B = -2 \\ z' = -x + y + y^2 \end{cases} \tag{3.38}$$

$$\begin{cases} x' = xy - Az \\ y' = x - y, \quad A = 1, B = 0.3 \\ z' = x + Bz \end{cases} \tag{3.39}$$

$$\begin{cases} x' = y + Az \\ y' = Bx^2 - y, \quad A = 3.9, B = 0.9 \\ z' = 1 - x \end{cases} \tag{3.40}$$

$$\begin{cases} x' = -z \\ y' = -x^2 - y, \quad A = 1.7, B = 1 \\ z' = A(1 + x) + y \end{cases} \tag{3.41}$$

$$\begin{cases} x' = Ay \\ y' = x + z^2, \quad A = -2, B = -2 \\ z' = 1 + y + Bz \end{cases} \tag{3.42}$$

$$\begin{cases} x' = y \\ y' = x - Az, \quad A = 1, B = 2.7 \\ z' = x + xz + By \end{cases} \tag{3.43}$$

$$\begin{cases} x' = Ay + z \\ y' = Bx + y^2, \quad A = 2.7, B = -1 \\ z' = x + y \end{cases} \tag{3.44}$$

$$\begin{cases} x' = -z \\ y' = x - y, \quad A = 3.1, B = 0.5 \\ z' = Ax + y^2 + Bz \end{cases} \tag{3.45}$$

$$\begin{cases} x' = A - y \\ y' = B + z, \quad A = 0.9, B = 0.4 \\ z' = xy - z \end{cases} \tag{3.46}$$

$$\begin{cases} x' = -x + Ay \\ y' = Bx + z^2, \quad A = -4, B = 1 \\ z' = 1 + x \end{cases} \tag{3.47}$$

## References


1. Brown, R. & Chua, L. O. [1996] "Clarifying chaos: Examples and counterexamples," *Inter. J. Bifurcation & Chaos* **6**(2), 219–249.
2. Bergé, P., Pomeau, Y. & Vidal, C. [1984] *Order within Chaos*, John-Wiley and Sons, New York.
3. Chua, L. O., Motomasa, K. & Takashi, M. [1986] "The double scroll family. I Rigorous proof of chaos," *IEEE Trans. Circuits and Systems* **33**, 1072–1097.
4. de Almeida, A. [1988] *Hamitonian Systems: Chaos and Quantization*, Cambridge University Press, Cambridge.
5. Devaney, R. L. [1989] *An introduction to chaotic Dynamical Systems*, Addison-Wesley, New York.
6. Ford, J. [1986] *Chaos: Solving the unsolvable, Predicting the unpredictable, from Chaotic Dynamics and Fractals*, Academic Press, New York.
7. Gulick, D. [1992] *Encounters with Chaos*, McGraw-Hill, New York.
8. Hassard, B., Zhang, J., Hastings, S. P. & Troy, W. C. [1994] "A computer proof that the Lorenz equations have chaotic solutions," *Applied Mathematics Letters* **7**(1), 79–83.
9. Hastings, S. P. & Troy, W. C. [1996] "A shooting approach to chaos in the Lorenz equations," *J. Differential Equations* **127**(1), 41–53.
10. Heidel, J. & Zhang F. [1999] "Nonchaotic behaviour in the three-dimensional quadratic systems II. The conservative case," *Nonlinearity* **12**, 617–633.
11. Heidel, J. & Zhang, F. [2007] "Nonchaotic and chaotic behaviour in three–dimensional quadratic systems: Five–one conservative cases," *Inter. J. Bifurcation & Chaos* **17**(6), 2049–2072.
12. Katok, A. [1980] "Lyapunov exponents, entropy and periodic points for diffeomorphisms," *Publ. Math. IHES* **51**,137–174.
13. Lorenz, N. E. [1963] "Deterministic non–periodic flow," *J. Atmos. Sci* **20**, 130–141.
14. Melasoma, J. M. [2009] "Non–chaotic behavior for a class of quadratic jerk equations," *Chaos, Solitons & Fractals* **39**, 533–539.
15. Rössler, J. C. [1976] "An equation for continuous chaos," *Phys. Lett. A* **57**, 397–398.
16. Shil'nikov, L. [1994] "Chua's circuit: Rigorous results and future problems," *Inter. J. Bifurcation & Chaos* **4**(3), 489–519.

17. Schuster, H. [1988] *Deterministic Chaos*, VCH, Weinheim, Germany.
18. Sprott, J. C. [1994] "Some simple chaotic flow," *Phys. Rev. E* **50**, 647–650.
19. Sprott, J. C. [1997] "Simplest dissipative chaotic flow," *Phys. Lett. A* **228**, 271–247.
20. Sprott, J. C. [2000a] "Simple chaotic systems and circuits," *Am J. Phys* **68**(8), 758–763.
21. Sprott, J. C. [2000b] "Algebraically simple chaotic flows," *Int. J. of Chaos theory and applications* **5**(2), 1–20.
22. Wiggins, S. [1992] *Chaotic Transport in Dynamical Systems*, Springer-Verlag, New York.
23. Zhang, F. & Heidel, J. [1997] "Nonchaotic behaviour in the three-dimensional quadratic systems," *Nonlinearity* **10**, 1289–1303.
24. Zhang, F., Heidel, J. & Le Borne, R. [2008] "Determining nonchaotic parameter Regions in some simple chaotic Jerk functions," *Chaos, Solitons & Fractals* **11**, 1413–1318.
25. Zhang, F. & Heidel. [2010] "Chaotic and nonchaotic behaviour in the three-dimensional quadratic systems: 5-1 Dissipative cases," Preprint.

# Chapter 4

# On Chaotic and Hyperchaotic Complex Nonlinear Dynamical Systems

Gamal M. Mahmoud

*Department of Mathematics, Faculty of Science*
*Assiut University, Assiut 71516, Egypt*
*E-mail: gmahmoud@aun.edu.eg, gmahmoud_56@yahoo.com*

Dynamical systems described by real and complex variables are currently one of the most popular areas of scientific research. These systems play an important role in several fields of physics, engineering, and computer sciences, for example, laser systems, control (or chaos suppression), secure communications, and information science. Dynamical basic properties, chaos (hyperchaos) synchronization, chaos control, and generating hyperchaotic behavior of these systems are briefly summarized. The main advantage of introducing complex variables is the reduction of phase space dimensions by a half. They are also used to describe and simulate the physics of detuned laser and thermal convection of liquid flows, where the electric field and the atomic polarization amplitudes are both complex. Clearly, if the variables of the system are complex the equations involve twice as many variables and control parameters, thus making it that much harder for a hostile agent to intercept and decipher the coded message. Chaotic and hyperchaotic complex systems are stated as examples. Finally there are many open problems in the study of chaotic and hyperchaotic complex nonlinear dynamical systems, which need further investigations. Some of these open problems are given.

**Keywords**: Complex, chaotic, hyperchaotic, attractors, periodic, dissipative, chaos, synchronization, control, Lyapunov exponents, stability analysis.

## Contents

## 4.1. Introduction

Chaotic and hyperchaotic complex nonlinear dynamical systems, where the main variables participating in the dynamics are real and complex, constitute fascinating developments in applied sciences. Nonlinear systems with one positive Lyapunov exponent are defined as chaotic systems and exhibit sensitive dependence on the initial conditions. A nonlinear dynamical system with more than one positive Lyapunov exponent is called a hyperchaotic system. Lyapunov exponents are a measure of the rate of divergence (or convergence) of two initially nearby orbits. The necessary conditions for a continuous-time autonomous system to be chaotic are to have three variables with at least one nonlinear term and the trivial fixed point is unstable. In 1997, Zhang and Jack proved that 3-dimensional (3D) dissipative quadratic systems of ordinary differential equations (ODE's) with a total of four terms on the right-hand side (RHS), can not exhibit chaos [Zhang & Heidel, 1997]. This result was proved, also, to 3D conservative quadratic systems [Heidel & Zhang, 1999; Yang & Chen, 2002; Lü *et al.*, 2004]. Later, it was known that chaotic behavior could be produced by a 3D quadratic autonomous system having five terms on the RHS, with at least one quadratic nonlinearity, or having six terms with a single quadratic nonlinearity [Sprott & Linz, 2000]. Sprott, in 1997, proposed the simplest example of a dissipative chaotic flow as:

$$\dddot{x} + A\ddot{x} - \dot{x}^2 + x = 0 \tag{4.1}$$

it can be equivalently written as three, first-order, ordinary differential equations with a total of five terms. It has only a single quadratic nonlinearity [Sprott, 1997]. This system exhibits a period-doubling route to chaos for $2.017 < A < 2.082$. In 1963, Lorenz found the first chaotic attractor in a 3D autonomous system when he studied the atmospheric convection [Lorenz, 1963]. This chaotic model is described by:

$$\dot{x} = \alpha(y - x), \quad \dot{y} = \gamma x - y - xz, \quad \dot{z} = xy - \beta z \tag{4.2}$$

For $a = 10, b = 8/3, c = 28$, this system has 2- scroll chaotic attractor. Also this system has three fixed points where are $(0,0,0), (\sqrt{\beta(\gamma-1)}, \sqrt{\beta(\gamma-1)}, \gamma-1), (-\sqrt{\beta(\gamma-1)}, -\sqrt{\beta(\gamma-1)}, \gamma-1)$.

Later, Rössler in 1976 constructed an even simpler 3D chaotic system, written as [Rössler, 1976]

$$x' = -(y+x),\ y' = x + ay,\ z' = zx - cz + b \tag{4.3}$$

which has, also, chaotic attractors when $a = b = 0.2$, and $c = 2$, 2.3, 3.5, 4.7, 5, 5.7, 6, 7, 8, 9, 10, 11. In 1979, he proposed further a real four-variables oscillator, i.e., hyperchaotic Rössler system, which can be described as follows [Rössler, 1979]:

$$\begin{cases} x' = by + cx \\ y' = 3 + yz \\ z' = -y - \omega \\ \omega' = x + z + a\omega \end{cases} \tag{4.4}$$

when $a = 0.25, b = 0.5$ and $c = 0.05$. This system exhibits a hyperchaotic behavior. This System displays hyperchaotic attractor when $a = 31$, $b = 2$, $c = 15$, $r = 2$. In 1999, Chen discovered a 3D chaotic system via a simple state feedback to the second equation in the Lorenz system, yielding [Chen & Ueta, 1999]:

$$\dot{x} = \alpha(y-x),\ \dot{y} = (\gamma - \alpha)x + cy - xz,\ \dot{z} = xy - \beta z \tag{4.5}$$

When $\alpha = 35, \beta = 3, \gamma = 28$, this system generate a 2-scroll chaotic attractor. This system has three fixed points: $(0,0,0)$, $(\sqrt{\beta\gamma}, \sqrt{\beta\gamma}, 2\gamma - \alpha)$, $(-\sqrt{\beta\gamma}, -\sqrt{\beta\gamma}, 2\gamma - \alpha)$.

Lü and Chen in 2002 reported a new a chaotic system which called by others Lu system.This system bridges the gap between the Lorenz system and the Chen system, which is: $x' = \rho\,(y - x)\,, y' = \nu y - xz, z' = xy - \mu z$. This system displays 2-scroll chaotic attractor when $\rho = 36, b = \mu, \nu = 20$ and has three fixed points $(0,0,0)\,, (\sqrt{\mu\nu}, \sqrt{\mu\nu}, \nu)$. See [Lü & Chen, 2002].

In a certain sense, Lü's equations represent a "transition" from the Lorenz attractor to the Chen attractor: If $a_{ij}$ are the elements of the $3 \times 3$ real matrix $A = (a_{ij})$ of the linearized equations about the trivial equilibrium point, the Lorenz system satisfies the condition $a_{12}a_{21} > 0$, Chen's system satisfies $a_{12}a_{21} < 0$, while for Lü's system one has $a_{12}a_{21} = 0$ [Vanecek & Celikovsky, 1996]. Thus, it may be said that the Lü attractor is "intermediate" between the Lorenz and Chen attractors. Lorenz, Chen and Lü systems are similar but nonequivalent.

1n 2003, Liu and Chen introduced a 3D continuous autonomous chaotic system, which can display complex 2- and 4-croll attractors [Liu & Chen, 2003]. This system can be described by the equations:

$$\dot{x} = a_1 x + b_1 yz, \dot{y} = a_2 y + b_2 xz, \dot{z} = a_3 z + b_3 xy \tag{4.6}$$

where $a_i$ and $b_i$, $i = 1, 2, 3$, are real constants. In 2004, Lü *et al.* introduced a 3D chaotic system as [Lü *et al.*, 2004]

$$\dot{x} = -[a_1 b_1/(a_1 + b_1)]x - yz + a_2, \dot{y} = a_1 y + xz, \dot{z} = b_1 z + xy \tag{4.7}$$

In 1994, Sprott performed an extensive computer search in which he found fourteen additional chaotic systems with six terms and one quadratic nonlinearity and five terms and two quadratic nonlinearities [Sprott, 1994]. A "*jerkfunction*" is a function $J$ such that the third-order ordinary differential equation can be written in the form [Gottlieb, 1996]:

$$\dddot{x} = J(\ddot{x}, \dot{x}, x) \tag{4.8}$$

where $J$ can be considered the time derivatives of an acceleration $\ddot{x}$. Linz [1997] showed that the Lorenz and the original Rössler models have rather complicated functional forms for $J$, but that Sprott's model $R$ can be written as [Sprott, 2000]

$$\dddot{x} = -\ddot{x} - 0.9x + x\dot{x} - 0.4 \tag{4.9}$$

A hyperjerk system is a dynamical system governed by an $n^{th}$ order differential equation with $n > 3$ describing the time evolution of a single scalar variable [Chlouverakis & Sprott, 2006; Linz, 2008].
In [Chen *et al.*, 2005] a system was introduced and studied, which is described by:

$$\dot{x} = a_1(y - x) + yz, \dot{y} = a_2 x - y - xz, \dot{z} = b_1 z + xy \tag{4.10}$$

and Qi *et al.* [2005] introduced a 4D autonomous system which is:

$$\begin{aligned} &\dot{x}_1 = a(x_2 - x_1) + x_2 x_3 x_4, \ \dot{x}_2 = b(x_1 + x_2) - x_1 x_3 x_4, \\ &\dot{x}_3 = c x_3 + x_1 x_2 x_4, \ \dot{x}_4 = -d a x_4 + x_1 x_2 x_3 \end{aligned} \tag{4.11}$$

where $a, b, c, d$ are positive constants parameters.
In 2009, Dadras and Momeni proposed a 3D chaotic system generating two, three, and four-scroll attractors:

$$\dot{x} = y - a_1 - x + b_1 yz, \dot{y} = a_2 y - xz + z, \dot{z} = a_3 xy - b_2 z \tag{4.12}$$

In the last four decades chaos has been studied within the science, mathematics, physics, and engineering. Different types of chaos and hyperchaos synchronization are introduced and studied, for example, complete synchronization (CS), anti-synchronization (AS), phase synchronization (PS), lag synchronization (LS), generalized synchronization (GS), projective and modified projective synchronization (PS and MPS) (e.g. [Qi & Chen, 2006;

Zhang & Lu, 2008; Femat & Solis-Perales, 2002; Femat *et al.*, 2000, 2005; Juan & Xing-yuan, 2007; Vincent & Laoye, 2007; Kim *et al.*, 2003, Liu *et al.*, 2006; Rosenblum *et al.*, 1997; Boccaletti *et al.*, 2002; Mainieri & Rehacek, 1999] and references therein). As a definition, we will refer to **chaos** (or **hyperchaos**) synchronization as a process wherein two chaotic (or hyperchaotic) systems (either identical or different) adjust a given property of their motion to a common behavior due to coupling or forcing. Complete synchronization is the strongest in the degree of correlation and describes the interaction of two identical systems, leading to their trajectories remaining identical in the course of temporal evolution, i.e., $x_d(t) = x_r(t)$, where $x_d(t)$ and $x_r(t)$ are the states of the drive and response systems, respectively. Generalized synchronization, as introduced for drive-response systems, is defined as the presence of a functional relationship between the states of the response and drive systems, i.e., $x_d(t) = f(x_r(t))$. Phase synchronization is the situation where two coupled hyperchaotic systems keep their phases in step with each other while their amplitudes remain uncorrelated. LS implies that the state variables of the two coupled systems become synchronized but with a time lag with respect to each other. Therefore, recently, much attention has been given to the LS, in which the state of the response system at time $t$ is asymptotically synchronous with the drive system at time $t - \tau$, namely, $\lim\limits_{t \longrightarrow \infty} |x_r(t) - x_d(t - \tau)| = 0$. Complete synchronization is special case of LS when $\tau = 0$. PS is a situation in which the state variables of the drive and response systems synchronize up to a constant scaling factor $\delta$. MPS is defined if the responses of the synchronized dynamical states synchronize up to a constant scaling matrix. Complete synchronization (CS) and anti-synchronization (AS) are special cases of PS where the scaling factor $\delta = 1$ and $-1$, respectively.

A wide variety of techniques has been employed to study chaos synchronization in chaotic systems of ODEs in real variables. Some of these are: Global synchronization, active control, adaptive control, linear and nonlinear control, feedback method and backstepping design [e.g., Agiza & Yassen, 2001; Elabbasy *et al.*, 2004; Chen & Lü, 2002; Ucar *et al.*, 2006; Yassen, 2006; Park, 2005a,b; Dai & Lonngren, 2000; Wu & Lü, 2003; Lei *et al.*, 2006; Huang *et al.*, 2004; Jiang *et al.*, 2003; Mahmoud *et al.*, 2007a]. In recent years synchronization of chaotic systems with different order and strictly different chaotic systems are studied (e.g.[ Femat *et al.*, 2002; Guan *et al.*, 2005; Bowong, 2004] and references therein). Synchronization of chaotic and hyperchaotic systems with unknown parameters are investigated (e.g.[Miao *et al.*, 2009; Sudheer & Sabir, 2009; Runzi, 2008; Yassen, 2006; Park, 2000a,b,c] and references therein). Chaos control and synchronization of Duffing, Duffing-Van der Pol and Van der Pol oscillators are studied (e.g. in [Barron &Sen, 2009, Bowong *et al.*, 2006; Kakmeni *et al.*, 2004, Mahmoud *et al.*, 2001b] and references therein).

In 1982, Fowler *et al.* introduced the complex Lorenz model as [Fowler *et al.*, 1982, 1983]:

$$\dot{x} = a(y - x), \quad \dot{y} = cx - y - xz, \quad \dot{z} = -bz + \frac{1}{2}(\bar{x}y + x\bar{y}) \tag{4.13}$$

where $x$ and $y$ are complex functions, z is a real function and $a, b$ and $c$ are positive parameters. Dots represent derivatives with respect to time and an overbar denotes complex conjugate function. These equations describe and simulate the physics of detuned lasers ([Rauh *et al.*, 1996] and references therein]). The functions $x, y$ and $z$ are related to the electric field and the atomic polarization amplitudes and the population inversion respectively. The basic properties of this system and equations of laser model are studied in [Mahmoud *et al.*, 2007c, 2009b].

Mahmoud [1998] has presented an approximate analytical method for finding periodic solutions of autonomous complex nonlinear dynamical systems of the form:

$$\ddot{z} + \omega^2 z + \epsilon f(z, \bar{z}, \dot{z}, \dot{\bar{z}}) = 0 \tag{4.14}$$

where $\epsilon$ is a small parameter, $z(t) = x(t) + iy(t)$ is a complex function, $i = \sqrt{-1}$ is the imaginary unit, $\bar{z}(t)$, is the complex conjugate function and $\omega^2$ is the constant frequency of the unperturbed ($\epsilon = 0$) oscillator, $f$ is an analytic function and dots represent derivatives with respect to time $\dot{z} = dz/dt$. Upon comparison with an alternative analytical method [Cveticanin, 2001, 1992a,b] it can be proved that the results of [Cveticanin, 1992a] are the zero-order results of the generalized averaging approach developed in [Mahmoud, 1998; Mahmoud & Bountis, 2004]. By parametrically excited complex dynamical systems, we mean systems which are governed by complex differential equations with periodic coefficients. Many dynamical systems of practical interest can be reduced to systems of this kind. In applied science there are a lot of problems described by two coupled nonlinear second order differential equations with periodic coefficients, which can be written using complex variables as [Mahmoud, 2000a,b,c, 2001a,b; Mahmoud & Bountis, 2004]:

$$\ddot{z} + \omega^2 z + \epsilon P(\Omega t) f(z, \bar{z}, \dot{z}, \dot{\bar{z}}) = 0 \tag{4.15}$$

where $P(\Omega t)$ is a periodic function with period $T = 2\pi/\Omega$.

Mahmoud and his co-authors studied different complex systems with random variables, e.g. stochastic Duffing and Duffing-Van der Pol systems [Xu *et al.*, 2005a, b, 2008].

In 2007, Mahmoud *et al.* introduced both complex Chen and Lü systems [Mahmoud *et al.*, 2007b, 2009a]. The complex Chen system expressed by:

$$\begin{aligned} \dot{x} &= \alpha\,(y-x) \\ \dot{y} &= (\gamma-\alpha)\,x - xz + \gamma y \\ \dot{z} &= 1/2\,(\bar{x}y + x\bar{y}) - \beta z \end{aligned} \tag{4.16}$$

while the complex Lü system written in the form:

$$\begin{aligned} \dot{x} &= \rho\,(y-x) \\ \dot{y} &= -xz + \nu y \\ \dot{z} &= 1/2\,(\bar{x}y + x\bar{y}) - \mu z \end{aligned} \tag{4.17}$$

where $\alpha, \beta, \gamma, \rho, \mu$ and $\nu$ are positive (real) parameters, $x = u_1 + iu_2$, $y = u_3 + iu_4$ are complex variables, $i = \sqrt{-1}$ and $u_i$, $z$ are real functions. Dots represent derivatives with respect to time and an "overbar" denotes complex conjugate variables. The complex Chen system (4.16) satisfies the condition $a_{24}a_{42} < 0$, while for the complex Lü system (4.17) we have $a_{24}a_{42} = 0$, where $A = (a_{ij})$ is the $5 \times 5$ matrix of the linear part of the equations about the origin $x = y = z = 0$. In 2008, Mahmoud presented a complex system (5D with real variables) which displays chaotic and hyperchaotic behaviors as [Mahmoud *et al.*, 2008b]:

$$\dot{x} = a(y-x) + yz, \quad \dot{y} = cx - y - xz, \quad \dot{z} = -bz + \frac{1}{2}(\bar{x}y + x\bar{y}) \tag{4.18}$$

where $x = v_1 + iv_2$, $y = v_3 + iv_4$ are complex functions (variables), $i = \sqrt{-1}$, $z = v_5$, and $v_i$, $i = 1, \ldots, 5$ are real functions.

More recently, Mahmoud *et al.* proposed several complex models for hyperchaotic Lorenz, Chen, and Lü systems using state feedback and complex periodic forcing [Mahmoud *et al.*, 2007d, 2008b,c, 2009c], for example,: Autonomous hyperchaotic Lorenz systems:

$$\begin{aligned} &\dot{x} = \alpha(y-x) + w, \; \dot{y} = \gamma x - y - xz \\ &\dot{z} = 1/2\,(\bar{x}y + x\bar{y}) - \beta z + w, \; \dot{w} = 1/2\,(\bar{x}y + x\bar{y}) - \sigma w \end{aligned} \tag{4.19}$$

where $w$ is a real variable and $\sigma$ is a control parameter. The dynamics of this system is more rich in the sense that exhibits both hyperchaotic and chaotic attractors as well as periodic, quasi-periodic (2-torus) and solutions that approach fixed points. And non-autonomous hyperchaotic Lorenz systems:

$$\dot{x} = \alpha(y-x) + k\exp(i\omega t), \; \dot{y} = \gamma x - y - xz, \; \dot{z} = 1/2\,(\bar{x}y + x\bar{y}) - \beta z \tag{4.20}$$

where $\omega$ and $k$ are positive parameters. The complex periodic control signal $k \exp(i\omega t)$ is added to the first equation of complex Lorenz system. This system is a 5D non-autonomous system and can be reduced to a 6D autonomous one by defining a new real variable $v_6 = \omega t$.

## 4.2. Examples

In this section we state chaotic and hyperchaotic complex examples which have been introduced and studied in our recent publications.

### 4.2.1. *Dynamical Properties of Chaotic Complex Chen System*

The complex Chen system expressed by: [Mahmoud *et al.*, 2007b]

$$\dot{x} = \alpha(y - x), \; \dot{y} = (\gamma - \alpha)x - xz + \gamma y, \; \dot{z} = 1/2(\bar{x}y + x\bar{y}) - \beta z \quad (4.21)$$

where $\alpha, \beta, \gamma$ are everywhere taken positive (real) parameters, $x = u_1 + iu_2$, $y = u_3 + iu_4$ are complex functions, $i = \sqrt{-1}$ and $z$ is a real function. In this subsection we state the basic dynamical analysis of our new system (4.21). The real version of (4.21) reads:

$$\begin{aligned} \dot{u}_1 &= \alpha(u_3 - u_1), \; \dot{u}_2 = \alpha(u_4 - u_2) \\ \dot{u}_3 &= (\gamma - \alpha)u_1 - u_1u_5 + \gamma u_3 \\ \dot{u}_4 &= (\gamma - \alpha)u_2 - u_2u_5 + \gamma u_4 \\ \dot{u}_5 &= u_1u_3 + u_2u_4 - \beta u_5 \end{aligned} \quad (4.22)$$

See Fig. 4.1. The complex Chen system (4.21) in the real version satisfies the condition $a_{13}a_{31} < 0$ (or $a_{24}a_{42} < 0$) where $A = (a_{ij})$ is the $5 \times 5$ matrix of the linear part of the equations (4.22). System (4.22) has the following basic dynamical properties:

**(1) Symmetry and invariance:** Symmetry about the $u_5$-axis, which is invariant for the coordinate transformation $(u_1, u_2, u_3, u_4, u_5) \longrightarrow (-u_1, -u_2, -u_3, -u_4, u_5)$.

**(2) Dissipation:** The divergence of the vector field $F$ of (4.22):

$$\nabla . F = \sum_{i=1}^{5} \frac{\partial \dot{u}_i}{\partial u_i} = 2\gamma - 2\alpha - \beta \quad (4.23)$$

so this system (4.22) is dissipative for the case:

$$\gamma < \alpha + \beta/2 \quad (4.24)$$

**(3) Equilibria and their stability:** The equilibria of this system are an isolated one at $E_0 = (0,0,0,0,0)$ (trivial fixed point) and a whole circle of equilibria given by the expression:

$$u_1^2 + u_2^2 = \beta(2\gamma - \alpha) \tag{4.25}$$

Trivial fixed point is stable if:

$$2\gamma < \alpha \ \ and \ \ \beta > 0$$

otherwise it is an unstable fixed point.
To study the stability of a whole *circle* of equilibria:
Setting $u_3 = u_1 = \sqrt{\beta(2\gamma-\alpha)}\cos\theta$ and $u_2 = u_4 = \sqrt{\beta(2\gamma-\alpha)}\sin\theta$, for $\theta \in [0, 2\pi]$. Therefore the nontrivial fixed points become:

$$\begin{aligned} E_\theta = (&\sqrt{\beta(2\gamma-\alpha)}\cos\theta,\ \sqrt{\beta(2\gamma-\alpha)}\sin\theta, \sqrt{\beta(2\gamma-\alpha)}\cos\theta \\ &, \sqrt{\beta(2\gamma-\alpha)}\sin\theta, 2\gamma-\alpha), \ \text{for } \theta \in [0, 2\pi] \end{aligned}$$

To study the stability of $E_\theta$ the Jacobian matrix of system (4.21) at $E_\theta$ is:

$$J_{E_\theta} = \begin{pmatrix} -\alpha & 0 & \alpha & 0 & 0 \\ 0 & -\alpha & 0 & \alpha & 0 \\ -\gamma & 0 & \gamma & 0 & -L_c \\ 0 & -\gamma & 0 & \gamma & -L_s \\ L_c & L_s & L_c & L_s & -\beta \end{pmatrix}$$

where $L_c = \sqrt{\beta(2\gamma-\alpha)}\cos\theta$ and $L_s = \sqrt{\beta(2\gamma-\alpha)}\sin\theta$.
Respectively, however, they have the same characteristic polynomial, which is:

$$\lambda(\lambda + \alpha - \gamma)(\lambda^3 + \lambda^2(\alpha + \beta - \gamma) + \gamma\beta\lambda - 2\alpha^2\beta + 4\alpha\gamma\beta) = 0$$

According to Routh-Hurwitz theorem the necessary and sufficient conditions for all the roots to have negative real parts if and only if:

$$\alpha > \gamma,\ \alpha > \gamma - \beta,\ \gamma > \alpha/2 \text{ and } \beta > \frac{2\alpha^2 + \gamma^2 - 5\alpha\gamma}{\gamma}$$

Otherwise they are unstable fixed points.
(4) **Lyapunov exponents**: For $\alpha = 42$, $\gamma = 26$, $\beta = 4$ we calculate the Lyapunov exponents as: $\lambda_1 = 1.29$, $\lambda_2 = 0$, $\lambda_3 = 0$, $\lambda_4 = -23.09$, $\lambda_5 = -30.14$. This means that our system (4.22) for this choice of $\alpha$, $\gamma$ and $\beta$ is a chaotic system since one of the Lyapunov exponents is positive. See Fig. 4.1c for maximum Lyapunov exponent.

**(5) Solutions of the complex Chen system:** The values of the parameters and the corresponding dynamical behaviors of (4.21) can be classified numerically for $\alpha = 42$, $\beta = 4$, as follows: (1) For $0 < \gamma < 21$, solutions of complex Chen system approach the trivial fixed point $(0, 0, 0, 0, 0)$, $(2\gamma \leqslant \alpha)$.
(2) For $21 < \gamma \leqslant 24$, solutions of complex Chen system approach one of the nontrivial fixed points, $(2\gamma > \alpha)$.
(3) For $24 < \gamma < 31$, complex Chen system has chaotic attractors.
(4) For $\gamma = 31$, this system has a periodic solution.
(5) For $31 < \gamma < 34$, complex Chen system has chaotic attractors.
(6) For $34 \leqslant \gamma \leqslant 40$, this system has periodic solutions.
Other values of $\alpha$, and $\beta$ are calculated as we did for $\gamma$ in Ref. [Mahmoud *et al.*, 2007b].

**(6) Synchronization of chaotic attractors:** Let us now study chaos synchronization of chaotic attractors of the complex Chen system (4.21) using the method of active control as follows: We assume that we have two complex Chen systems and denote the drive system by the subscript 1, while the response system to be controlled is denoted by the subscript 2. The drive and response systems are thus defined respectively as:

$$\begin{aligned} \dot{x}_1 &= \alpha\,(y_1 - x_1) \\ \dot{y}_1 &= (\gamma - \alpha)\,x_1 - x_1 z_1 + \gamma y_1 \\ \dot{z}_1 &= 1/2\,(\bar{x}_1 y_1 + x_1 \bar{y}_1) - \beta z_1 \end{aligned} \tag{4.26}$$

and

$$\begin{aligned} \dot{x}_2 &= \alpha\,(y_2 - x_2) + (v_1 + i v_2) \\ \dot{y}_2 &= (\gamma - \alpha)\,x_2 - x_2 z_2 + \gamma y_2 + (v_3 + i v_4) \\ \dot{z}_2 &= 1/2\,(\bar{x}_2 y_2 + x_2 \bar{y}_2) - \beta z_2 + v_5 \end{aligned} \tag{4.27}$$

where $x_1 = u_{11} + i u_{21}$, $y_1 = u_{31} + i u_{41}$, $z_1 = u_{51}$, $x_2 = u_{12} + i u_{22}$, $y_2 = u_{32} + i u_{42}$ and $z_2 = u_{52}$, "overbar" denotes complex conjugation, $v_1 + i v_2$, $v_3 + i v_4$ and $v_5$ are complex and real control functions respectively, which are to be determined and all $u_{ij}$ and $v_i$ variables are real. The complex system (4.26) can thus be written in the form of five real first order ODEs:

$$\begin{aligned} \dot{u}_{11} &= \alpha\,(u_{31} - u_{11}) \\ \dot{u}_{21} &= \alpha\,(u_{41} - u_{21}) \\ \dot{u}_{31} &= (\gamma - \alpha)\,u_{11} - u_{11} u_{51} + \gamma u_{31} \\ \dot{u}_{41} &= (\gamma - \alpha)\,u_{21} - u_{21} u_{51} + \gamma u_{41} \\ \dot{u}_{51} &= u_{11} u_{31} + u_{21} u_{41} - \beta u_{51} \end{aligned} \tag{4.28}$$

In order to obtain the active control signals, we define as the "errors" between the drive and the response system the quantities:

$$\begin{aligned} e_{u_1} + ie_{u_2} &= x_2 - x_1 = (u_{12} - u_{11}) + i(u_{22} - u_{21}) \\ e_{u_3} + ie_{u_4} &= y_2 - y_1 = (u_{32} - u_{31}) + i(u_{42} - u_{41}) \\ e_{u_5} &= z_2 - z_1 = u_{52} - u_{51} \end{aligned} \tag{4.29}$$

Subtracting (4.26) from (4.27) and using (4.29) we get:

$$\begin{aligned} \dot{e}_{u_1} + i\dot{e}_{u_2} &= \alpha\left[(e_{u_3} - e_{u_1}) + i\,(e_{u_4} - e_{u_2})\right] + (v_1 + iv_2) \\ \dot{e}_{u_3} + i\dot{e}_{u_4} &= -\alpha\,(e_{u_1} + ie_{u_2}) + \gamma\left[(e_{u_1} + e_{u_3}) + i(e_{u_2} + e_{u_4})\right] \\ &\quad - e_{u_5}\,(u_{12} + iu_{22}) - u_{51}\,(e_{u_1} + ie_{u_2}) + (v_3 + iv_4) \\ \dot{e}_{u_5} &= -\beta e_{u_5} + u_{11}e_{u_3} + u_{32}e_{u_1} + u_{21}e_{u_4} + u_{42}e_{u_2} + v_5 \end{aligned} \tag{4.30}$$

Eq. (4.30) describes a dynamical system via which the "errors" $e_{u_i}$ evolve in time and its ODEs, when separated in real and imaginary parts, become:

$$\begin{aligned} \dot{e}_{u_1} &= \alpha\,(e_{u_3} - e_{u_1}) + v_1 \\ \dot{e}_{u_2} &= \alpha\,(e_{u_4} - e_{u_2}) + v_2 \\ \dot{e}_{u_3} &= -\alpha e_{u_1} + \gamma(e_{u_1} + e_{u_3}) - e_{u_5}u_{12} - u_{51}e_{u_1} + v_3 \\ \dot{e}_{u_4} &= -\alpha e_{u_2} + \gamma(e_{u_2} + e_{u_4}) - e_{u_5}u_{22} - u_{51}e_{u_2} + v_4 \\ \dot{e}_{u_5} &= -\beta e_{u_5} + u_{11}e_{u_3} + u_{32}e_{u_1} + u_{21}e_{u_4} + u_{42}e_{u_2} + v_5 \end{aligned} \tag{4.31}$$

For positive parameters $\gamma$, $\alpha$ and $\beta$, we may now define a Lyapunov function for this system by the following positive definite quantity:

$$V\,(t) = 1/2\sum_{i=1}^{5} e_{u_i}^2 \tag{4.32}$$

Note now that the total time derivative of $V(t)$ along the solutions of system (15) is:

$$\begin{aligned} \dot{V}(t) = &-(\alpha e_{u_1}^2 + \alpha e_{u_2}^2 + \beta e_{u_5}^2) \\ &+ e_{u_3}\left[\gamma(e_{u_1} + e_{u_3}) - e_{u_5}u_{12} - u_{51}e_{u_1}\right] \\ &+\; e_{u_4}\left[\gamma(e_{u_2} + e_{u_4}) - e_{u_5}u_{22} - u_{51}e_{u_2}\right] \\ &+ e_{u_5}\left[u_{11}e_{u_3} + u_{32}e_{u_1} + u_{21}e_{u_4} + u_{42}e_{u_2}\right] + \textstyle\sum_{i=1}^{5} v_i e_{u_i} \end{aligned} \tag{4.33}$$

Thus, if we choose active control functions $v_i$ such that:

$$\begin{aligned} v_1 &= 0 \\ v_2 &= 0 \\ v_3 &= -\left[\gamma(e_{u_1} + e_{u_3}) - e_{u_5}u_{12} - u_{51}e_{u_1}\right] \\ v_4 &= -\left[\gamma(e_{u_2} + e_{u_4}) - e_{u_5}u_{22} - u_{51}e_{u_2}\right] \\ v_5 &= -\left[u_{11}e_{u_3} + u_{32}e_{u_1} + u_{21}e_{u_4} + u_{42}e_{u_2}\right] \end{aligned} \tag{4.34}$$

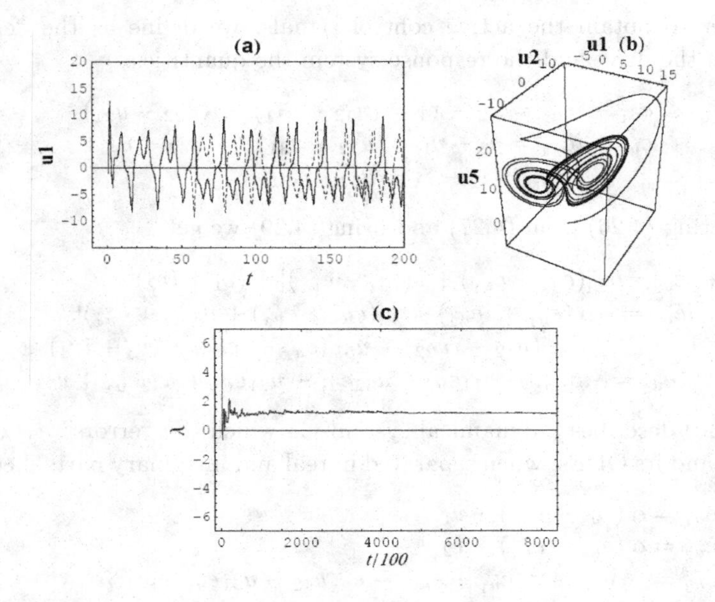


Fig. 4.1. (a) Two numerically calculated solutions of (4.22) for $\alpha = 42$, $\beta = 4$ and $\gamma = 26$ with $t_0 = 0$, $u_1(0) = -1$, $u_2(0) = 2$, $u_3(0) = 3, u_4(0) = 4$, $u_5(0) = -5$ (solid curve) and $u_1(0) = -.999$, $u_2(0) = 2$, $u_3(0) = 3.001$, $u_4(0) = 4$, $u_5(0) = -5$ (dotted curve). Note the exponential separation that becomes evident at $t \cong 55$, indicating the chaotic nature of the orbits. (b) A chaotic attractor of (4.22) at $\alpha = 42$, $\beta = 4$ and $\gamma = 26$ in $(u_1, u_3, u_5)$ space and the same initial conditions leading to the solid curve in (a). (c) The Maximum Lyapunov Exponent of this attractor is clearly positive, indicating that the motion on the attractor is chaotic for the same values of parameters and initial conditions as in (b).

Eq. (4.33) yields:

$$\dot{V}(t) = -(\alpha e_{u_1}^2 + \alpha e_{u_2}^2 + \beta e_{u_5}^2) < 0 \tag{4.35}$$

Since $V(t)$ is positive definite and its derivative is negative definite, Lyapunov's direct method implies that the equilibrium point $e_{u_i} = 0, i = 1, ..., 5$ of the system (4.31) is asymptotically stable, which means that $e_{u_i} \longrightarrow 0$ as $t \longrightarrow \infty$, for all $i = 1, 2, ..., 5$. Systems (4.26) and (4.27) with (4.34) are solved numerically, for $\alpha = 42$, $\beta = 4$ and $\gamma = 26$ and initial conditions $u_{11}(0) = -1$, $u_{21}(0) = 2$, $u_{31}(0) = 3$, $u_{41}(0) = 4$, $u_{51}(0) = -5$ and $u_{12}(0) = -13$, $u_{22}(0) = -12$, $u_{32}(0) = -13$, $u_{42}(0) = -14$, $u_{52}(0) = 40$. The results are depicted in Figs. 4.2 and 4.3. In Fig. 4.2 the solutions of (4.26) and (4.27) are plotted subject to different initial conditions and show that chaos synchronization is indeed achieved after a very small interval in

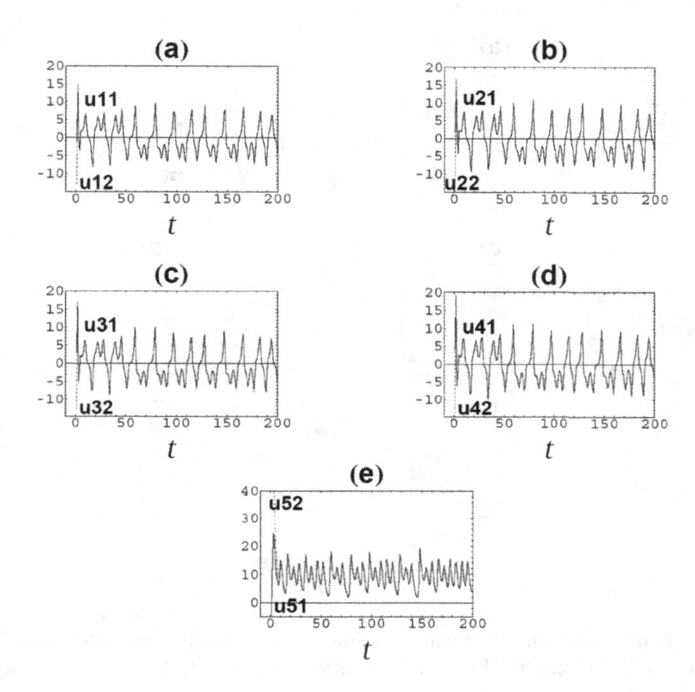


Fig. 4.2. Chaos synchronization of systems (4.26) and (4.27) with (4.34) for $\alpha = 42$, $\beta = 4$ and $\gamma = 26$ with $t_0 = 0$, $u_{11}(0) = -1$, $u_{21}(0) = 2$, $u_{31}(0) = 3$, $u_{41}(0) = 4$, $u_{51}(0) = -5$ and $u_{12}(0) = -13$, $u_{22}(0) = -12$, $u_{32}(0) = -13$, $u_{42}(0) = -14$, $u_{52}(0) = 40$. (a) $u_{11}(t)$ and $u_{12}(t)$ versus $t$, (b) $u_{21}(t)$ and $u_{22}(t)$ versus $t$, (c) $u_{31}(t)$ and $u_{32}(t)$ versus $t$, (d) $u_{41}(t)$ and $u_{42}(t)$ versus $t$, (e) $u_{51}(t)$ and $u_{52}(t)$ versus $t$.

time $t$. In Fig. 4.3, on the other hand, it can be seen that the synchronization errors $e_{u_i}$ converge to zero, as expected from the above analytical considerations.

### 4.2.2. *Hyperchaotic Complex Lorenz Systems*

Recently, we introduced and analyzed new hyperchaotic complex Lorenz systems [Mahmoud *et al.*, 2008c]. These systems are 6-dimensional systems of real first order autonomous differential equations and their dynamics are very complicated and more rich. In this study we extended the idea of adding state feedback control and introduce the complex periodic forces to generate hyperchaotic behaviors.

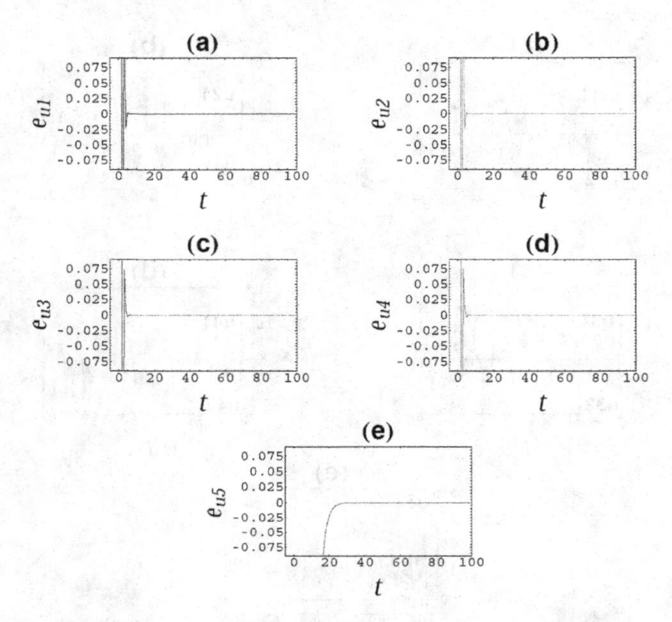


Fig. 4.3. Time evolution of synchronization errors, obtained as solutions of system (4.31). (a) $(e_{u_1}, t)$ diagram, (b) $(e_{u_2}, t)$ diagram, (c) $(e_{u_3}, t)$ diagram, (d) $(e_{u_4}, t)$ diagram, (e) $(e_{u_5}, t)$ diagram.

#### 4.2.2.1. *Autonomous Hyperchaotic Complex Lorenz Systems*

We added the state feedback control to the first and the third equations of complex Lorenz system as:

$$\begin{aligned} \dot{x} &= \alpha(y-x)+w,\ \dot{y}=\gamma x-y-xz \\ \dot{z} &= 1/2\,(\bar{x}y+x\bar{y})-\beta z+w,\ \dot{w}=1/2\,(\bar{x}y+x\bar{y})-\sigma w \end{aligned} \tag{4.36}$$

where $w$ is a real variable and $\sigma$ is a control parameter. The dynamics of (4.36) is more rich in the sense that exhibits both hyperchaotic and chaotic attractors as well as periodic, quasi-periodic(2-torus) and solutions that approach fixed points. It is complicated since (4.36) has different attractors for very small interval values of the parameter $\alpha, \beta, \gamma$ and $\sigma$. In this part we present the complex behaviors of our new system (4.36). The real version of (4.36) reads:

$$\begin{aligned} \dot{v}_1 &= \alpha\,(v_3-v_1)+v_6,\ \dot{v}_2=\alpha\,(v_4-v_2) \\ \dot{v}_3 &= \gamma v_1-v_3-v_1v_5,\ \dot{v}_4=\gamma v_2-v_4-v_2v_5 \\ \dot{v}_5 &= v_1v_3+v_2v_4-\beta v_5+v_6,\ \dot{v}_6=v_1v_3+v_2v_4-\sigma v_6 \end{aligned} \tag{4.37}$$

System (4.37) has the following basic dynamical properties:

**(1) Symmetry and invariance:** From (4.37),we note that this system is invariant under the transformation $(v_1, v_2, v_3, v_4, v_5, v_6) \longrightarrow (v_1, -v_2, v_3, -v_4, v_5, v_6)$.
Therefore, if $(v_1, v_2, v_3, v_4, v_5, v_6)$ is a solution of (4.37), then $(v_1, -v_2, v_3, -v_4, v_5, v_6)$ is also a solution of the same system.

**(2) Dissipation:** System (4.37) is dissipative under the condition $2\alpha + \beta + \sigma + 2 > 0$ since:

$$\frac{\partial \dot{v}_1}{\partial v_1} + \frac{\partial \dot{v}_2}{\partial v_2} + \ldots + \frac{\partial \dot{v}_6}{\partial v_6} = -(2\alpha + \beta + \sigma + 2) \tag{4.38}$$

**(3) Fixed points of (4.37):** The equilibria of system (4.37) can be calculated by solving the following system of equations $\dot{v}_1 = 0$, $\dot{v}_2 = 0$, $\dot{v}_3 = 0$, $\dot{v}_4 = 0$, $\dot{v}_5 = 0$ and $\dot{v}_6 = 0$ to get three isolated equilibria $E_0 = (0, 0, 0, 0, 0, 0)$, $E_1 = (v_1^*, 0, v_3^*, 0, v_5^*, v_6^*)$ and $E_2 = (v_1^{**}, 0, v_3^{**}, 0, v_5^{**}, v_6^{**})$ where:

$$\begin{aligned}
v_1^* &= \frac{\beta[2\sigma\alpha^2\beta(1+\sigma)(\gamma-1) - \beta^2\gamma + s_1]}{\alpha(1+\sigma)[-\beta^2(2\gamma-1) + s_1]} \\
v_3^* &= \frac{\alpha\sigma[-\beta^2(2\gamma-1) + s_1]}{\beta[2\sigma\alpha^2(1+\sigma) + \beta]} \\
v_5^* &= \frac{2\sigma\alpha^2\beta(1+\sigma)(\gamma-1) - \beta^2\gamma + s_1}{\beta[2\sigma\alpha^2(1+\sigma) + \beta]} \\
v_6^* &= \frac{2\sigma\alpha^2\beta(1+\sigma)(\gamma-1) - \beta^2\gamma + s_1}{(1+\sigma)[2\sigma\alpha^2(1+\sigma) + \beta]} \\
v_1^{**} &= \frac{\beta[2\sigma\alpha^2\beta(1+\sigma)(\gamma-1) - \beta^2\gamma - s_1]}{\alpha(1+\sigma)[-\beta^2(2\gamma-1) - s_1]} \\
v_3^{**} &= \frac{\alpha\sigma[-\beta^2(2\gamma-1) - s_1]}{\beta[2\sigma\alpha^2(1+\sigma) + \beta]}, \\
v_5^{**} &= \frac{2\sigma\alpha^2\beta(1+\sigma)(\gamma-1) - \beta^2\gamma - s_1}{\beta[2\sigma\alpha^2(1+\sigma) + \beta]} \\
v_6^{**} &= \frac{2\sigma\alpha^2\beta(1+\sigma)(\gamma-1) - \beta^2\gamma - s_1}{(1+\sigma)[2\sigma\alpha^2(1+\sigma) + \beta]}
\end{aligned} \tag{4.39}$$

and

$$s_1 = \sqrt{\beta^4\gamma^2 - 4\sigma\alpha^2\beta^3(1+\sigma)(\gamma-1)(2\gamma-1)}$$

To study the stability of $E_0$, the Jacobian matrix of system (4.37) is:

$$L_{i,j} = \begin{pmatrix} -\alpha & 0 & \alpha & 0 & 0 & 1 \\ 0 & -\alpha & 0 & \alpha & 0 & 0 \\ \gamma - v_5 & 0 & -1 & 0 & -v_1 & 0 \\ 0 & \gamma - v_5 & 0 & -1 & -v_2 & 0 \\ v_3 & v_4 & v_1 & v_2 & -\beta & 1 \\ v_3 & v_4 & v_1 & v_2 & 0 & -\sigma \end{pmatrix}$$

The characteristic polynomial of $L_{i,j}$ at the equilibrium $E_0$ is:

$$(\lambda + \sigma)(\lambda + \beta)(\lambda^2 + \lambda(\alpha + 1) + \alpha(1 - \gamma))^2 = 0 \quad (4.40)$$

Then the eigenvalues are:

$$\begin{aligned} &\lambda_1 = -\sigma,\ \lambda_2 = -\beta \\ &\lambda_3 = \lambda_4 = -\frac{1}{2}\left[\ (\alpha + 1) + \sqrt{(\alpha + 1)^2 + 4\alpha(\gamma - 1)}\ \right] \\ &\lambda_5 = \lambda_6 = -\frac{1}{2}\left[\ (\alpha + 1) - \sqrt{(\alpha + 1)^2 + 4\alpha(\gamma - 1)}\ \right] \end{aligned} \quad (4.41)$$

So this fixed point is **stable** if $\sigma$, $\beta$, $\alpha$ are positive and $0 < \gamma < 1$. Otherwise it is an **unstable** fixed point.

**(4) Lyapunov exponents and dimensions:** We calculated both Lyapunov exponents of our system (4.37) and Lyapunov dimension of its attractors. System (4.37) in vector notation can be written as:

$$\dot{V}\,(t) = h\,(V(t); \eta) \quad (4.42)$$

where $V(t) = [v_1(t), v_2(t), v_3(t), v_4(t), v_5(t), v_6(t)]^t$ is the state space vector, $h = [h_1, h_2, h_3, h_4, h_5, h_6]^t$, $\eta$ is a set of parameters and $[...]^t$ denoting transpose. The equations for small deviations $\delta V$ from the trajectory $V(t)$ are:

$$\delta\dot{V}(t) = L_{ij}(V(t); \eta)\delta V, \qquad i, j = 1, 2, 3, 4, 5, 6 \quad (4.43)$$

where $L_{i,j} = \dfrac{\partial h_i}{\partial v_j}$ is the Jacobian matrix. The Lyapunov exponents $\lambda_i$ of the system is defined by:

$$\lambda_i = \lim_{t \longrightarrow \infty} \frac{1}{t} \log \frac{\|\delta v_i(t)\|}{\|\delta v_i(0)\|} \quad (4.44)$$

The signs of Lyapunov exponents provide a good classification of the dynamics of our system (4.37). The attractors of (4.37) can be described as: (+, +, -, -, -, -) or (+, +, 0, -, -, -) a hyperchaotic attractor. The Lyapunov dimension of the attractors of (4.37) according to Kaplan-Yorke conjecture is defined as:

$$D = M + \frac{\sum_{i=1}^{M} \lambda_i}{|\lambda_{M+1}|} \tag{4.45}$$

such that $M$ the largest integer for which $\sum_{i=1}^{M} \lambda_i > 0$ and $\sum_{i=1}^{M+1} \lambda_i < 0$. To find $\lambda_i$, Eqs. (4.42) and (4.43) must be numerically solved simultaneously. For the case $\alpha = 14$, $\beta = 5$, $\gamma = 40$ and $\sigma = 15$ with the initial conditions $t_0 = 0$, $v_1(0) = 1$, $v_2(0) = 2$, $v_3(0) = 3, v_4(0) = 4$, $v_5(0) = 5$ and $v_6(0) = 6$ we calculate the Lyapunov exponents as: $\lambda_1 = 1.83561$, $\lambda_2 = 0.12738$, $\lambda_3 = 0$, $\lambda_4 = -8.58376$, $\lambda_5 = -21.72827$, $\lambda_6 = -29.35575$. This means that our system (4.37) for this choice of $\alpha, \gamma$, $\beta$ and $\sigma$ is a hyperchaotic system since $\lambda_1$ and $\lambda_2$ are positive values and dissipative system since the sum of Lyapunov exponents is negative. The Lyapunov dimension of this hyperchaotic attractor using Eq (4.45) is $D \cong 3.22868$.

**(5) Different forms of HCLS's:** We constructed different versions of the hyperchaotic complex nonlinear system based on complex Lorenz system by adding a state feedback controller to different equations of this system. By adding $w(1+i)$ to the first equation of complex Lorenz system we get the system:

$$\begin{aligned} \dot{x} &= \alpha\,(y-x) + w(1+i),\ \dot{y} = \gamma x - y - xz \\ \dot{z} &= 1/2\,(\bar{x}y + x\bar{y}) - \beta z\ ,\dot{w} = 1/2\,(\bar{x}y + x\bar{y}) - \sigma w \end{aligned} \tag{4.46}$$

For the case $\alpha = 14$, $\beta = 5$, $\gamma = 45$ and $\sigma = 5.5$ and with the same above initial conditions we calculate the Lyapunov exponents using Eq. (4.44) as: $\lambda_1 = 1.85633$, $\lambda_2 = 0.27995$, $\lambda_3 = 0$, $\lambda_4 = -7.9393$, $\lambda_5 = -21.950$, $\lambda_6 = -30.67027$, and its Lyapunov dimension is $D \cong 3.26907$. Therefore system (4.46) has hyperchaotic behavior since $\lambda_1$ and $\lambda_2$ are positive. The third system has hyperchaotic behavior can be considered as:

$$\begin{aligned} \dot{x} &= \alpha\,(y-x) + w,\ \dot{y} = \gamma x - y - xz + w \\ \dot{z} &= 1/2\,(\bar{x}y + x\bar{y}) - \beta z,\ \dot{w} = 1/2\,(\bar{x}y + x\bar{y}) - \sigma w \end{aligned} \tag{4.47}$$

The Lyapunov exponents for the above system for $\alpha = 14$, $\beta = 5$, $\gamma = 45$ and $\sigma = 15$ are:
$\lambda_1 = 1.64816$, $\lambda_2 = 0.44169$, $\lambda_3 = 0$, $\lambda_4 = -21.70814$, $\lambda_5 = -22.78434$, $\lambda_6 = -29.7280$, and $D \cong 3.09627$. A fourth system is described by:

$$\begin{aligned} \dot{x} &= \alpha\,(y-x) + iw,\ \dot{y} = \gamma x - y - xz + iw \\ \dot{z} &= 1/2\,(\bar{x}y + x\bar{y}) - \beta z,\ \dot{w} = 1/2\,(\bar{x}y + x\bar{y}) - \sigma w \end{aligned} \tag{4.48}$$

From Eq. (4.44) and (4.45) for $\alpha = 14$, $\beta = 5$, $\gamma = 40$ and $\sigma = 13$ we get: $\lambda_1 = 1.50033$, $\lambda_2 = 0.41015$, $\lambda_3 = 0$, $\lambda_4 = -18.8073$, $\lambda_5 = -22.61082$, $\lambda_6 = -29.73889$, and the Lyapunov dimension of this hyperchaotic attractor is $D \cong 3.10158$. One constructs the hyperchaotic complex Lorenz system as:

$$\begin{aligned} \dot{x} &= \alpha\,(y - x)\,, \; \dot{y} = \gamma x - y - xz + w \\ \dot{z} &= 1/2\,(\bar{x}y + x\bar{y}) - \beta z + w, \; \dot{w} = 1/2\,(\bar{x}y + x\bar{y}) - \sigma w \end{aligned} \tag{4.49}$$

Calculating Eq. (4.44) and (4.45), one obtains for: $\alpha = 14$, $\beta = 6.28$, $\gamma = 45$ and $\sigma = 20$, the Lyapunov exponents are: $\lambda_1 = 1.56388$, $\lambda_2 = 0.53336$, $\lambda_3 = 0$, $\lambda_4 = -23.05613$, $\lambda_5 = -28.98295$, $\lambda_6 = -31.24986$, $D \cong 3.09096$. The last system we can constructed it is:

$$\begin{aligned} \dot{x} &= \alpha\,(y - x) + iw, \; \dot{y} = \gamma x - y - xz + w \\ \dot{z} &= 1/2\,(\bar{x}y + x\bar{y}) - \beta z, \; \dot{w} = 1/2\,(\bar{x}y + x\bar{y}) - \sigma w \end{aligned} \tag{4.50}$$

For the case $\alpha = 14$, $\beta = 5$, $\gamma = 40$ and $\sigma = 25$ we obtain: $\lambda_1 = 1.48302$, $\lambda_2 = 0.20773$, $\lambda_3 = 0$, $\lambda_4 = -22.2823$, $\lambda_5 = -30.66714$, $\lambda_6 = -35.29904$, $D \cong 3.075878$. It is clear that all the above systems have two positive Lyapunov exponents and their sum is negative, which means that they are hyperchaotic and dissipative systems. The basic properties of systems (4.46),...,(4.50) can be similarly treated as we did for the system (4.37).

#### 4.2.2.2. *Non-Autonomous Hyperchaotic Complex Lorenz System*

The idea of introducing a complex periodic forcing is applied to complex Lorenz system to generate a hyperchaotic complex Lorenz system as follows:

$$\dot{x} = \alpha(y - x) + k\exp(i\omega t), \; \dot{y} = \gamma x - y - xz, \; \dot{z} = 1/2\,(\bar{x}y + x\bar{y}) - \beta z \tag{4.51}$$

where $\omega$ and $k$ are positive parameters. The complex periodic control signal $k\exp(i\omega t)$ is added to the first equation of complex Lorenz system. System (4.51) is a 5D non-autonomous system and can be reduced to a 6D autonomous one by defining a new real variable $v_6 = \omega t$. The real version of (4.51) with $v_6 = \omega t$ reads:

$$\begin{aligned} \dot{v}_1 &= \alpha\,(v_3 - v_1) + k\cos(v_6), \; \dot{v}_2 = \alpha\,(v_4 - v_2) + k\sin(v_6) \\ \dot{v}_3 &= \gamma v_1 - v_3 - v_1 v_5, \; \dot{v}_4 = \gamma v_2 - v_4 - v_2 v_5, \\ \dot{v}_5 &= v_1 v_3 + v_2 v_4 - \beta v_5, \; \dot{v}_6 = \omega \end{aligned} \tag{4.52}$$

We note that this system is dissipative under the condition $2\alpha + \beta + 2 > 0$. For the case $\alpha = 15$, $\beta = 5$, $\gamma = 45$, $\omega = 13$ and $k = 10$ and at the same initial conditions of Fig. 4.1 the Lyapunov exponents for system (4.52) are

$\lambda_1 = 1.50902$, $\lambda_2 = 0.45679$, $\lambda_3 = 0$, $\lambda_4 = -23.3619$, $\lambda_5 = -31.97240$, $\lambda_6 = 0$.
This means that our system (4.52) for this choice of $\alpha, \gamma, \beta$, $\omega$ and $k$ is a hyperchaotic one since it has two positive Lyapunov exponents. The Lyapunov dimension of this hyperchaotic attractor using Eq. (4.45) is $D \cong 4.08414$. Based on Lyapunov exponents $\lambda_i$, Eq. (4.44) we calculate the parameters values of system (4.52) at which, chaotic, hyperchaotic, periodic and quasi-periodic attractors exist. System (4.52) does not have fixed points since $\dot{v}_6 = \omega \neq 0$. **Fix** $\alpha = 15$, $\beta = 5$, $\gamma = 45$, $k = 10$, **and vary** $\omega$ (1) our system (4.52) has hyperchaotic attractors for $\omega \in [10.90, 15.3]$, and $[20.64, 21.06]$ (2) The chaotic attractors are exist for $\omega \in [0, 8.14]$, $[8.27, 10.90]$, $[15.03, 20.64]$, and $[23.59, 200]$. (3) The quasi-periodic solutions (2-torus) are exist for $\omega$ lies in $[8.20, 8.23]$, and $[8.24, 8.27]$ This system has also, different solutions for very small interval values of $\omega$, for example, has hyperchaotic attractors for $\omega \in (21.21, 21.23]$, chaotic attractors for $\omega \in (21.12, 21.21]$, and quasi-periodic solutions t for $\omega \in (14.46, 14.47]$.

**Other HCLS's using complex periodic forcing:** We constructed different forms (or versions) of the hyperchaotic complex Lorenz systems by introducing complex periodic forces. One of them is our system (4.52). Another system to generate hyperchaotic behavior is:

$$\dot{x} = \alpha(y - x), \ \dot{y} = \gamma x - y - xz + k \exp(i\omega t), \ \dot{z} = 1/2\,(\bar{x}y + x\bar{y}) - \beta z \quad (4.53)$$

For the case $\alpha = 15$, $\beta = 5$, $\gamma = 45$, $\omega = 11$ and $k = 10$ and at the same initial conditions of Fig. 4.1 the Lyapunov exponents are: $\lambda_1 = 1.50349$, $\lambda_2 = 0.41403$, $\lambda_3 = 0$, $\lambda_4 = -23.55422$, $\lambda_5 = -31.7391$, $\lambda_6 = 0$, $D \cong 4.08140$. The third system can be considered as:

$$\dot{x} = \alpha(y - x) + k(1 + i)\cos(\omega t), \ \dot{y} = \gamma x - y - xz, \ \dot{z} = 1/2\,(\bar{x}y + x\bar{y}) - \beta z \quad (4.54)$$

Using Eq. (4.44) for the case $\alpha = 15$, $\beta = 5$, $\gamma = 45$, $\omega = 14$ and $k = 10$ the Lyapunov exponents of (4.54) are:
$\lambda_1 = 2.08118$, $\lambda_2 = 0.15098$, $\lambda_3 = 0$, $\lambda_4 = -23.09784$, $\lambda_5 = -32.50937$, $\lambda_6 = 0$, $D \cong 4.096637$. A fourth form is described by:

$$\dot{x} = \alpha(y - x), \ \dot{y} = \gamma x - y - xz + k(1 + i)\cos(\omega t), \ \dot{z} = 1/2\,(\bar{x}y + x\bar{y}) - \beta z \quad (4.55)$$

The Lyapunov exponents of (4.55) at the same parameter values of (4.54) are: $\lambda_1 = 1.88780$, $\lambda_2 = 0.30321$, $\lambda_3 = 0$, $\lambda_4 = -23.10793$, $\lambda_5 = -32.46331$, $\lambda_6 = 0$, and the Lyapunov dimension is $D \cong 4.094816$. One constructs the hyperchaotic Lorenz system as:

$$\dot{x} = \alpha(y - x) + k(1 + i)\sin(\omega t), \ \dot{y} = \gamma x - y - xz, \ \dot{z} = 1/2\,(\bar{x}y + x\bar{y}) - \beta z \quad (4.56)$$

As we did for system (4.55) and using the same parameter values the Lyapunov exponents and Lyapunov dimension are: $\lambda_1 = 2.11361$, $\lambda_2 = 0.13294$, $\lambda_3 = 0$, $\lambda_4 = -23.04253$, $\lambda_5 = -32.53083$, $\lambda_6 = 0$, $D \cong 4.09749$. The last system can be considered as:

$$\dot{x} = \alpha(y - x), \ \dot{y} = \gamma x - y - xz + k(1 + i)\sin(\omega t), \ \dot{z} = 1/2\,(\bar{x}y + x\bar{y}) - \beta z \quad (4.57)$$

For the case $\alpha = 15$, $\beta = 5$, $\gamma = 45$, $\omega = 15$ and $k = 10$ the Lyapunov exponents and D are: $\lambda_1 = 1.82750$, $\lambda_2 = 0.34214$, $\lambda_3 = 0$, $\lambda_4 = -23.18547$, $\lambda_5 = -32.2315$, $\lambda_6 = 0$, $D \cong 4.093577$. All the above systems have two positive Lyapunov exponents and their sum is negative. The dynamics of systems (4.53),...,(4.57) can be similarly studied as we did for (4.52).

## 4.3. Open Problems

This section contains many open problems in the study of chaotic and hyperchaotic complex nonlinear dynamical systems, which need further investigations. Some of these open problems are:

**Problem 4.1.** *Different kinds of synchronization and control of two completely different chaotic complex nonlinear systems, e.g. complex Duffing and Van der Pol systems.*

**Problem 4.2.** *Control and different types of synchronization of two identical (or non-identical) hyperchaotic complex systems, e.g. hyperchaotic Lorenz, Chen, and Lü systems.*

**Problem 4.3.** *Basic properties, control and synchronization of two identical (or non-identical) stochastic hyperchaotic complex systems.*

**Problem 4.4.** *Design electronic circuits for chaotic and hyperchaotic complex nonlinear dynamical systems.*

**Problem 4.5.** *Hyperchaotic complex nonlinear dynamical systems with unknown parameters.*

**Problem 4.6.** *Stochastic chaotic and hyperchaotic complex systems with unknown parameters (e.g. stochastic hyperchaotic complex Lorenz).*

**Problem 4.7.** *Fractional order chaotic and hyperchaotic complex systems (e.g. stochastic chaotic complex Duffing).*

**Problem 4.8.** *Piece wise-linear and nonlinear chaotic and hyperchaotic complex nonlinear dynamical systems.*

**Problem 4.9.** *Modified chaotic and hyperchaotic complex nonlinear systems (e.g. modified complex Chen).*

**Problem 4.10.** *Non-smooth chaotic and hyperchaotic complex nonlinear systems.*

**Problem 4.11.** *Stochastic Non-smooth chaotic and hyperchaotic complex nonlinear systems.*

**Problem 4.12.** *Complex Chua and modified Chua systems.*

**Problem 4.13.** *Complex jerk, hyperjerk, and Sprott systems.*

**Problem 4.14.** *Stability analysis, synchronization and control of chaotic and hyperchaotic complex nonlinear systems with different order.*

**Problem 4.15.** *Chaotic and hyperchaotic complex nonlinear systems with different order and unknown parameters.*

**Problem 4.16.** *Higher order chaotic and hyperchaotic complex nonlinear systems (more than 6D).*

**Problem 4.17.** *Generating chaotic and hyperchaotic attractors with two, three and more-scroll of complex nonlinear systems.*

## 4.4. Conclusions

The dynamics of chaotic and hyperchaotic complex nonlinear systems continues to be a challenging field for many researchers,concentrating on chaos and hyperchaos synchronization, stability analysis, bifurcation phenomena, and chaos (or hyperchaos) control. In this paper, we have sought to demonstrate that equally interesting problems arise and with broad applications, when the dynamics is expressed in terms of complex variables. In the literature much attention has been devoted to nonlinear dynamical systems with real variables. As is well-known, there exist interesting cases of dynamical systems where the main variables participating in the dynamics are complex, as for example when amplitudes of electromagnetic fields and atomic polarization are involved [Rauh *et al.*, 1996; Fowler *et al.*, 1982, 1983]. Introducing complex variables (or increasing the dimension) is also crucial in chaos synchronization used in secure communications, where one wishes to maximize the content and security of the transmitted information. Some of our proposed open problems become reasonable material for Ph.D. theses. It is hoped that the results reviewed here and the proposed open problems increase our knowledge of the dynamics of chaotic and hyperchaotic complex nonlinear dynamical systems, which is still far from what has been achieved to date for dynamical systems with real variables.

## References


1. Agiza, H. N. & Yassen, M. T. [2001] "Synchronization systems of Rössler and Chen chaotic dynamical systems using active control," *Phys. Lett. A* **278**, 191–197.
2. Barron, M. A. & Sen, M. [2009] "Synchronization of four coupled Van der Pol oscillators," *Nonlinear Dyn* **56**, 357–367.
3. Boccaletti, S., Kurth, J., Osipov, G., Valladaras, D. L. & Zhou, C. S. [2002] "The synchronization of chaotic systems," *Phys. Rep* **366**, 1–101.
4. Bowong, S., Kakmeni, F. M. M. & Dimi, J. L. [2006] "Chaos control in uncertain Duffing oscillator," *J. Sound. Vib* **292**, 869–880.
5. Bowong, S. [2004] "Stability analysis for the synchronization of chaotic systems with different order: application to secure communications," *Phys. Lett. A* **326**, 102–113.
6. Chen, G. & Ueta, T. [1999] "Yet another chaotic attractor," *Inter. J. Bifurcation & Chaos* **9**(7), 1465–1466.
7. Chen, G., Du, S., Chen, Z. & Yan, Z. [2005] "Analysis of a new chaotic system," *Phys. A* **352**, 295–308.
8. Chen, S. & Lü, J. [2002] "Synchronization of an uncertain chaotic system via adaptive control," *Chaos, Solitons & Fractals* **14**, 643–647.
9. Chlouverakis, K. E. & Sprott, J. C. [2006] "Chaotic hyperjerk systems," *Chaos, Solitons & Fractals* **28**, 739–746.
10. Cveticanin, L. [2001] "Analytical approach for the solution of the complex valued strongly nonlinear differential equation of Duffing type," *Phys. A* **297**, 348–360.
11. Cveticanin, L. [1992a] "Approximate analytical solutions to a class of nonlinear equations with complex functions," *J. Sound. Vib* **157**, 289–302.
12. Cveticanin, L. [1992b] "Approximate solutions for a system of two coupled differential equations," *J. Sound. Vib* **152**(2), 375–380.
13. Dadras, S. & Momeni, H. R. [2009] "A novel three-dimensional autonomous chaotic system generating two, three and four-scroll attractors," *Phys. Lett. A* **373**, 3637–3642.
14. Dai, E. W. & Lonngren, K. E. [2000] "Sequential synchronization of two Lorenz systems using active control," *Chaos, Solitons & Fractals* **11**, 1041–1044.

15. Elabbasy, E. H., Agiza, H. N. & El-Dessoky, M. [2004] "Synchronization of modified Chen system," *Inter. J. Bifurcation & Chaos* **14**(11), 3969–3979.
16. Femat, R., Kocarev, van Gerven, L. & Monsivais-Perez, M. E. [2005] "Towards generalized synchronization of strictly different chaotic systems," *Phys. Lett. A* **342**, 247–255.
17. Femat, R. & Solis-Perales, G. [1999] "On the chaos synchronization phenomena," *Phys. Lett. A* **262**, 50–60.
18. Femat, R., Alvarez-Rmirez, J. & Fernadez-Anaya, G. [2000] "Adaptive synchronization of higer-order chaotic systems: A feedback with low-order parametrization," *Phys. D* **139**, 231–246.
19. Femat, R. & Solis-Perales, G. [2002] "Synchronization of chaotic systems with different order," *Phys. Rev. E* **65**, 036226.
20. Fowler, A. C., Gibbon, J. D. & McGuinnes, M. T. [1983] "The real and complex Lorenz equations and their relevance to physical systems," *Phys. D* **7**, 126–134.
21. Fowler, A. C., McGuinness, M. J. & Gibbon, J. D. [1982] "The complex Lorenz equations," *Phys. D* **4**(2), 139–163.
22. Gottlieb, H. P. W. [1996] "Question 38. What is the simplest jerk function that gives chaos?," *Am. J. Phys* **64**, 525.
23. Guan, S., Lai, C. H. & Wei, G. W. [2005] "Phase synchronization between two essentially different chaotic systems," *Phys. Rev. E* **72**, 016205.
24. Huang, L., Feng, L. R. & Wang, M. [2004] "Synchronization of chaotic systems via nonlinear control," *Phys. Lett. A* **320**, 271–275.
25. Heidel, J. & Zhang, F. [1999] "Nonchaotic behavior in three-dimensional quadratic systems II. The conservative case," *Nonlinearity* **12**, 617–633.
26. Jiang, G. P., Tang, K. S. & Chen, G. [2003] "A simple global synchronization criterion for coupled chaotic systems," *Chaos, Solitons & Fractals* **15**, 925–935.
27. Juan, M. & Xing-yuan, W. [2007] "Nonlinear observer based phase synchronization of chaotic systems," *Phys. Lett. A* **369**, 294–298.
28. Kakmeni, F. M. M., Bowong, S., Tchawoua, C. & Kaptouom, E. [2004] "Strange attractors and chaos control in a Duffing-Van der Pol oscillator with two external periodic forces," *J. Sound. Vib* **277**, 783–799.
29. Kim, C. M., Rim, S., Kye, W. H., Ryu, J. W. & Park, Y. J. [2003] "Anti-synchronization of chaotic oscillators," *Phys. Lett. A* **320**, 39–46.
30. Lei, Y., Xu, W., Shen, J. & Fang, T. [2006] "Global synchronization of two parametrically excited systems using active control," *Chaos, Solitons & Fractals* **28**, 428–436.
31. Linz, S. J. [1997] "Nonlinear dynamical models and jerky motion," *Am. J. Phys* **65**, 523–526.
32. Linz, S. J. [2008] "On hyperjerky systems," *Chaos, Solitons & Fractals* **37**, 741–747.
33. Liu, W. B. & Chen, G. [2003] "A new chaotic system and its generation," *Inter. J. Bifurcation & Chaos* **12**, 261–267.
34. Liu, W. B., Xiao, J., Qian, X. & Yang, J. [2006] "Antiphase synchronization in coupled chaotic oscillators," *Phys. Rev. E* **73**, 057203.

35. Lorenz, E. N. [1963] "Deterministic non-periodic flows," *J. Atoms. Sci* **20**(1), 130–141.
36. Lü, J., Chen, G. & Cheng, D. [2004] "A new chaotic system and beyond: The generalized Lorenz-like system," *Inter. J. Bifurcation & Chaos* **14**, 1507–1537.
37. Lü, J. & Chen, G. [2002] "A new chaotic attractor coined," *Inter. J. Bifurcation & Chaos* **12**(3), 659–661.
38. Mahmoud, G. M., Aly, S. A. & Farghaly, A. A. [2007a] "On chaos synchronization of a complex two coupled dynamos system," *Chaos, Solitons & Fractals* **33**(1), 178–187.
39. Mahmoud, G. M. [1998] "Approximate solutions of a class of complex nonlinear dynamical systems," *Phys. A* **253**, 211–222.
40. Mahmoud, G. M. & Bountis, T. [2004] "The dynamics of systems of complex nonlinear oscillators: A Review," *Inter. J. Bifurcation & Chaos* **14**(11), 3821–3846.
41. Mahmoud, G. M. [2001a] "A theorem for n-dimensional strongly nonlinear dynamical systems," *Int. J. Non-Linear Mech* **36**, 1013–1018.
42. Mahmoud, G. M., Mohamed, A. A. & Aly, S. A. [2001b] "Strange attractors and chaos control in periodically forced complex Duffing's oscillators," *Phys. A* **292**,193–206.
43. Mahmoud, G. M. & Aly, S. A. [2000a] "Periodic attractors of complex damped nonlinear systems," *Int. J. Non-Linear Mech* **35**, 309–323.
44. Mahmoud, G. M. & Aly, S. A. [2000b] "On periodic solutions of parametrically excited complex nonlinear dynamical systems," *Phys. A* **278**, 390–404.
45. Mahmoud, G. M. & Bountis, T. [1988] "Synchronized periodic solutions of a class of periodically driven nonlinear oscillators," *J. Applied Mech* **110**(55), 721–728.
46. Mahmoud, G. M., Bountis, T. & Ahmed, S. A. [2000c] "Stability analysis for systems of nonlinear Hill's equations," *Phys. A* **286**, 133–146.
47. Mahmoud, G. M., Bountis, T. & Mahmoud, E. E. [2007b] "Active control and global synchronization of complex Chen and Lü systems," *Inter. J. Bifurcation & Chaos* **17**(12), 4295–4308.
48. Mahmoud, G. M., Al–Kashif, M. A. & Aly, S. A. [2007c] "Basic properties and chaotic synchronization of complex Lorenz system," *Int. J. Mod. Phys. C*, **18**(2), 253–265.
49. Mahmoud, G. M., Mahmoud, E. E. & Ahmed. M. E. [2007d] "A hyperchaotic complex Chen system and its dynamics," *Int. J. Appl. Math. Stat* **12**(D07), 90–100.
50. Mahmoud, G. M., Bountis, T., AbdEl-Latif, G. M., & Mahmoud, E. E. [2009a] "Chaos synchronization of two different chaotic complex Chen and Lü systems," *Nonlinear Dyn* **55(**1-2), 43–53.
51. Mahmoud, G. M., Bountis, T., Al-Kashif, M. A., & Aly, S. A. [2009b] "Dynamical properties and synchronization of complex nonlinear equations for detuned lasers," *Dyn. Systems* **24**(1), 63–79.
52. Mahmoud, G. M., Mahmoud, E. E & Ahmed. M. E. [2009c] "On the hyperchaotic complex Lü system," *Nonlinear. Dyn* **58**, 725–738.

53. Mahmoud, G. M., Aly, S. A. & Al-Kashif, M. A. [2008a] "Dynamical properties and chaos synchronization of a new chaotic complex nonlinear system," *Nonlinear. Dyn* **51**(1-2) 171–181.
54. Mahmoud, G. M., Al-Kashif, M. A. & Farghaly, A. A. [2008b] "Chaotic and hyperchaotic attractors of a complex nonlinear system," *J. Phys. A: Math. Theor* **41**(5), 055104.
55. Mahmoud, G. M., Ahmed, M. E., & Mahmoud, E. E. [2008c] "Analysis of hyperchaotic complex Lorenz systems," *Int. J. Modern Physics. C* **19**(10), 1477–1494.
56. Mainieri, R. & Rehacek, J. [1999] "Projective synchronization in three-dimensional chaotic systems," *Phys. Rev. Lett* **82**, 3042–3045.
57. Miao, Q., Tang, Y., Lu, S. & Fang, J. [2009] "Lag synchronization of a class of chaotic with unknown parameters," *Nonlinear. Dyn* **57**, 107–112.
58. Park, J. H. [2005a] "Chaos synchronization of a chaotic system via nonlinear control," *Chaos, Solitons & Fractals* **25**(3), 579–584.
59. Park, J. H. [2005b] "On synchronization of unified chaotic systems via nonlinear control," *Chaos, Solitons & Fractals* **25**(2), 699–704.
60. Park, J. H. [2005c] "Adaptive synchronization of hyperchaotic Chen system with unknown parameters," *Chaos, Solitons & Fractals* **26**, 959–964.
61. Qi, G. & Chen, G. [2006] "Analysis and circuit implementation of a new 4D chaotic system," *Phys. Lett. A* **352**, 386–397.
62. Qi, G., Du, S., Chen, G. & Chen, Z. [2005] "On a four-dimensional chaotic system," *Chaos, Solitons & Fractals* **23**, 1671–1682.
63. Rauh, A., Hannibal, L. & Abraham, N. [1996] "Global stability properties of the complex Lorenz model," *Phys. D* **99**, 45–58
64. Rosenblum, M., Pikovsky, A. & Kurths, J. [1997] "From phase to lag synchronization in coupled chaotic oscillators," *Phys. Rev. Lett* **78**, 4193–4196.
65. Rössler, O. E. [1979] "An equation for hyperchaos," *Phys. Lett. A* **71**, 155–157.
66. Rössler, O. E. [1976] "An equation for continuous chaos," *Phys. Lett. A* **57**, 397–398.
67. Runzi, L. [2008] "Adaptive function projective synchronization of Rössler hyperchaotic system," *Phys. Lett. A* **372**, 3667–3671.
68. Sprott, J. C. & Linz, S. J. [2000] "Algebraically simple chaotic flows," *Int. J. Chaos TH. Appl.* **5**, 3–22.
69. Sprott, J. C. [1994] "Some simple chaotic flows," *Phys. Rev. Lett. E* **50**, R647–R650.
70. Sprott, J. C. [1997] "Simplest dissipative chaotic flow," *Phys. Lett. A* **228**, 271–274.
71. Sprott, J. C. [2000] "Simple chaotic systems and circuits," *Am. J. Phys* **68**, 758–763.
72. Sudheer, K. S. & Sabir, M. [2009] "Adaptive modified function projective synchronization between hyperchaotic Lorenz system and hyperchaotic system with uncertian parameters," *Phys. Lett. A* **373**, 3743–3748.

73. Ucar, A., Lonngren, K. E. & Bai, E. W. [2006] "Synchronization of the unified chaotic systems via active control," *Chaos, Solitons & Fractals* **27**(5), 1292–1297.
74. Vanecek, A. & Celikovsky, S. [1996] *Control Systems: From Linear Analysis to Synthesis of Chaos,* Prentice-Hall, London.
75. Vincent, U. E. & Laoye, J. A. [2007] "Synchronization, anti-synchronization and current transports in non-identical chaotic ratchets," *Phys. A* **384**, 230–240.
76. Wu, X. & Lü, J. [2003] "Parameter identification and backstepping control of uncertain Lü system," *Chaos, Solitons & Fractals* **18**, 721–729.
77. Xu, Y., Xu, W. & Mahmoud, G. M. [2005a] "Generating chaotic limit cycles for a complex Duffing–Van der Pol system with random phase," *Int. J. Modern Physics C* **16**(9), 1437–1447.
78. Xu, Y., Xu, W., Mahmoud, G. M. & Lei, Y. [2005b] "Beam-beam interaction models under narrow-band random excitation," *Phys. A* **346**(3–4), 372–386.
79. Xu, Y., Xu, W. & Mahmoud, G. M. [2008] "On a Complex Duffing systems with random excitation," *Chaos, Solitons & Fractals* **35**(1), 126–132.
80. Yang, X. S. & Chen, G. [2002] "On non-chaotic behavior of a class of jerky systems," *Far East J. Dyn. Syst* **4**, 27–38.
81. Yassen, M. T. [2006] "Adaptive chaos control and synchronization of uncertain new chaotic dynamical system," *Phys. Lett, A* **350**, 36–43.
82. Zhang, F. & Heidel, J. [1997] "Non-chaotic behavior in three-dimensional quadratic systems," *Nonlinearty* **10**, 1289–1303.
83. Zhang, Q. & Lu, J. A. [2008] "Full state hybrid lag projective synchronization in chaotic (hyperchaotic) systems," *Phys. Lett. A* **372**, 1416–1421.

# Chapter 5

# On the Study of Chaotic Systems with Non-Horseshoe Template

Anirban Ray[1], A. Roy Chowdhury[2] and Sankar Basak[3]

[1,2]*High Energy Physics Division, Department of Physics*
*Jadavpur University, Calcutta - 700032, India*
[2]*E-mail: asesh_r@yahoo.com*

[3]*Physics Department, Shyambazar A.V. School, Calcutta - 700002, India*

Relevance of electronic circuits in the study of nonlinear dynamical systems is now a proven fact. After the advent of Chua's Circuit, there has been tremendous development, and subsequent research have seen to be highly fruitful. In this paper, we have developed a new chaotic system with the help of an electronic circuit which was made from usual electronic components, and analyzed its subsequent output. The corresponding time series and attractor are studied from the topological point of view. It is observed that the Poincaré map leads to the binary symbolic representation of the kneading sequence and the attractor is populated by quite a high number of period 2, 3 and 4 orbits. Later the linking number and the torsion of the system are computed, and the template is designed leading to a global understanding of chaotic scenario.

**Keywords**: New chaotic attractor, time series analysis, symbolic dynamics, template structure.

## Contents

## 5.1. Introduction

Application of electronic circuit to the study of nonlinear dynamical system and analyzed of a set of nonlinear ODE are vice-versa. While the former approach gives an experimental visualization of the chaotic output of the time series obtainable from the non-linear ODEs, the later approach can be used as a tool for the analysis of the circuit whose output we usually observe on an oscilloscope. Perhaps one of the most famous system falling in this category is Chua's circuit [Matsumoto, 1984]. Chua proposed this as a good paradigm for a novel nonlinear system, realizable practically in the laboratory. Among various possible utilities of such realization of non-linear system in terms of circuits [Freire *et al.*, 1993 ], the most important one is the effective understanding of synchronization of two such systems [Cuomo *et al.*, 1993] and their subsequent use in encryption for message transmission. Since such an event is basically dependent on the chaotic aspect of the corresponding dynamical system, it is very much pertinent to have deep understanding of the organization of the unstable periodic orbits of such a nonlinear system. At the present moment such an understanding can only be realized through the topological analysis of the attractor and the symbolic encoding of periodic orbits. A major amount of research in this direction was done by Birman and Williams, Gilmore, Lefranc and many others [Birman *et al.*, 1983; Solari *et. al.*, 1988; Lefranc *et al.*, 1993; Letellier *et. al.* 1994; Letellier *et. al.* 1995; Gilmore *et al.*, 1995; Gilmore *et. al.*, 2002]. Mindlin was the first one to apply the present method to experimental data series [Mindlin *et al.*, 1991; Mindlin *et al.*, 1992]. Most of the case the horseshoe structure plays prominent role in many attractors. Attractors with non-horseshoe structure are rare but not unheard off [Boulant *et al.*, 1997; Boulant *et al.*, 1998]. Recent works on the same field, which must be noted, are [Kiers *et al.*, 2004; Ray *et al.*, 2009].

So here in this paper we have proposed a new system developed from an electronic circuit involving the usual operational amplifiers and multiplexer, along with the standard set of resistor and capacitors which ultimately leads to a new nonlinear system. The electronic simulation of the circuit shows a very beautiful chaotic attractor, along with some of its periodic orbits. Next we deal with the details of the bifurcation pattern and its various properties. The overall global behavior is then studied through a detailed study of unstable periodic orbits and their organizational pattern. In this respect we have given the detailed symbolic sequence of them. Which are then used to find the torsion and linking number associated with them. Lastly the template or knot holder is constructed.

## 5.2. Formulation

The electronic circuit under consideration is shown in Fig. 5.1. In it we have used seven op-amps IC's UA741 and the multiplexer AD 633. The output is observed in the double beam Oscilloscope (Scientific SM410). The values of the other electronic components are as shown in the Fig. 5.1. Here all resistance are of five percent tolerance.

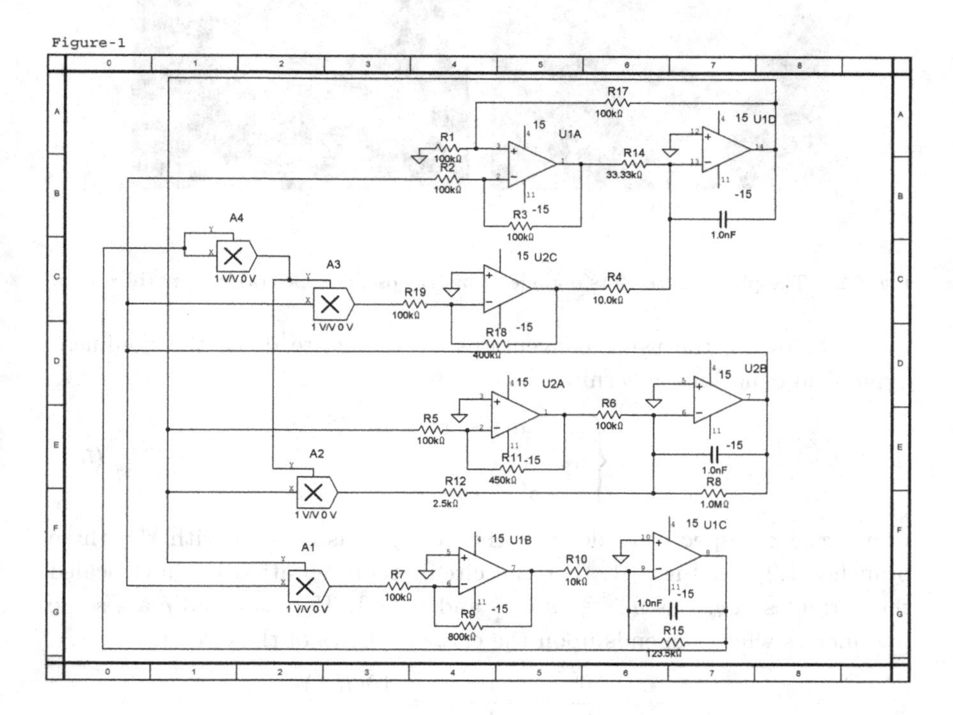


Fig. 5.1. Circuit diagram of the electronic circuit.

In Fig. 5.2 we show a snapshot of the attractor as seen on the oscilloscope screen. Fig. 5.2a shows the $xz$ projection where as Fig. 5.2b shows the $yz$ projection as obtained in oscilloscope. The triggering of the circuit is done by the following conditions noted below.

(a) To keep the number of active components to a minimum to reduce the extraneous noise introduced by the circuit.

(b) Voltage limit for the various components are not to be exceeded to ensure that the components operate within manufacturer specifications.

(c) It should be noted that at all point of the circuit the signal is lager than the background noise.

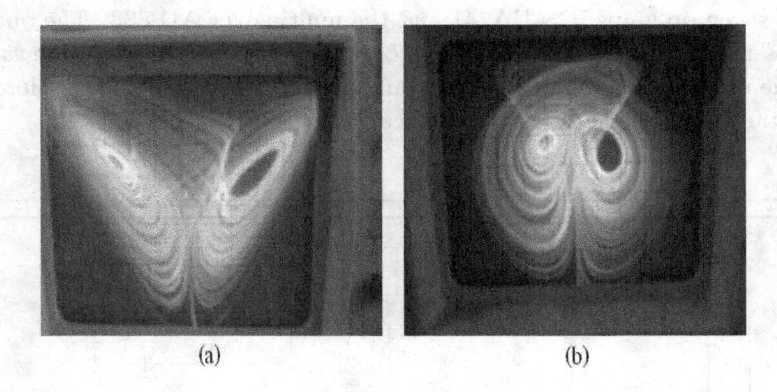

Fig. 5.2. The phase space plots obtained from the oscilloscope. (a) $x$ vs $z$. (b) $y$ vs $z$.

By following the usual conventions we can write down the nonlinear dynamical equations governing the circuit as:

$$\begin{cases} x' = a(y-x) + yz^2 \\ y' = cx - y - xz^2 \\ z' = xy - bz \end{cases} \tag{5.1}$$

where $x, y, z$ respectively denotes the voltages associated with the three branches 1,2,3 in Fig. 5.1. For the circuit representation we have scaled the variables $x, y, z$ as $u = \frac{x}{4}$, $v = \frac{y}{4}$ and $w = \frac{z}{2}$. Here $a, b$ and $c$ are some parameters which depends upon the characteristics of the circuit.

$$a = \frac{10^6}{R_{14}}; b = \frac{10^6}{R_{15}}; c = \frac{10(R_{11})}{R_5}$$

This is nonlinear dynamical system of third order which is introduced for the first time.

The fixed points are given as

$$E_0(0,0,0), E_1(x_+, y_+, z_+), E_2(x_+, y_-, z_+)$$
$$E_3(x_-, y_+, z_+), E_4(x_-, y_-, z_+)$$

where

$$x_{\pm} = \pm\sqrt{\frac{bz}{c - z^2}}$$

$$y_{\pm} = \pm\sqrt{bz(c - z^2)}$$

$$z_{\pm} = \pm\sqrt{\frac{c - a \pm \sqrt{(c-a)^2 - 4a(1-c)}}{2}}$$

Linearizing about $E_0$, we get the characteristics equations;

$$\lambda^3 + (a + b + 1)\lambda^2 + b(a + 1)\lambda = 0 \tag{5.2}$$

with eigenvalues:

$$\lambda_1 = 0, \lambda_2 = -(a + 1), \lambda_3 = -b$$

So this equilibrium $E_0$ is always stable. Same procedure is applied to the point $E(x_0, y_0, z_0)$ leads to,

$$\lambda^3 + A_1\lambda^2 + A_2\lambda + A_3 = 0$$

where the coefficients are given below. Since this equation contains even powers of $x_0$ and $y_0$ so that stability depends only on the sign of $z_0$.

$$\begin{cases} A_1 = a + b + 1 \\ A_2 = ({z_0}^2 + a)({z_0}^2 - c) + a + b + ab + 2z_0(x_0^2 - y_0^2) \\ A_3 = a(b + 2x_0^2 z_0) + \xi \\ \xi = (a + z_0^2)(bz_0^2 - bc + 2x_0y_0z_0) - 2x_0y_0z_0(c - z_0^2) - 2y_0^2 z_0 \end{cases} \tag{5.3}$$

The roots of Eq. (5.2) gives the indication of various types of stability, corresponding to the perturbation around $(x_0, y_0, z_0)$. Now Routh-Hurwitz criterion dictates that the real parts of these roots are positive if and only if $A_1 > 0, A_3 > 0$ and $A_1A_2 > A_3$.

For our analysis we have restricted ourselves to the values of $a \in [15, 35], b = 8.1$ and $c = 45.0$. For these set we can also have a pair of complex conjugate eigenvalues leading to a Hopf bifurcation. In this case, our fixed point is $(0, 0, 0)$. But we do not study these routes. On the other hand, in Fig. 5.3 we show bifurcation diagram for $a \in [15, 35]$ and the above mentioned values of b, c. This was drawn from the data collected from the electronic circuit shown in Fig. 5.1. From Fig. 5.3 it is quite clear that the system gives full blown chaos at $a = 25.0, b = 8.1$ and $c = 45.0$. At these parameter values, the data are analyzed in the following section.

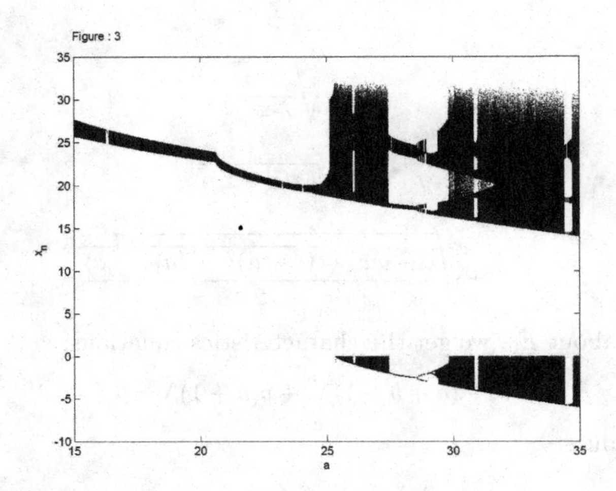


Fig. 5.3. The bifurcation diagram for $a \in [15.0, 35.0]$, $b = 8.1$ and $c = 45.0$.

## 5.3. Topological Analysis and Its Invariants

Before proceeding to the actual analysis, it is important to say few words about the necessity and relevance of the present method in the domain of nonlinear dynamics. In fact, main idea of such an analysis is organizing host of stable and unstable orbits residing in an attractor to study their geometry and get a global picture.

If one wants to understand chaos in a dynamical system, one must able to visualize underlying structure of a strange attractor. In particular, we must understand the organization of different periodic orbits in a strange attractor. The usual quantitative measures, which are generally used, can not do this job. One such aspect is the different linking of the orbits usually expressed through Gaussian linking numbers and another aspect is the relative rotation rate, which is introduced by Gilmore *et al.* Of course, one should remember that many of these concepts are at present applicable for three dimensional systems only.

In fact, the different stretching and squeezing mechanisms generate different types of strange attractor. The topological analysis of the unstable periodic orbits provides a clear fingerprint for these mechanisms. There is a geometric structure that supports all the unstable periodic orbits with the same unique organization. These organized structure are called a knot holder or branched manifolds or templates. They are identified by a set of integers.

Furthermore, there is a straightforward way of extracting the signature of a strange attractor from the data. The input is a time series, while the output consists of a branched manifold. The intrinsic idea of a topological analysis is to reveal the ways of stretching and squeezing mechanisms and their relations to the geometry of the attractor. The basic idea in its most elementary form is the Smale horseshoe. Because it is the main underlying structure of many of the chaotic systems. Of course there exists other variations which called reverse horseshoe, non horseshoe, etc.

The third mechanism, which is very crucial in the formation of an attractor, is the tearing mechanism which plays a vital role in the Lorenz system. The whole idea of a topological analysis is the study of these procedures (stretching, squeezing and tearing) for determining the geometry of an attractor and analyzing their variations with parameters. Such a phenomenon is usually known as unfolding of an attractor. If the control parameters are changed new unstable periodic orbits arises but the underlying manifold remains the same.

A basic tool for performing the above analysis is the so called symbolic dynamics, which provides a simple but faithful representation of the chaotic dynamical system. Quite often it is obtained by computing the Poincaré section of the attractor and symbolic dynamics give a very interesting view of the various mechanisms which is underlying a given dynamical system by simplifying the time series. In this connection, one may note that unstable periodic orbits can be located by the method of closed return segments which can be extracted from the data. The next important tool is the Birman–William theorem. This theorem actually deals with the fact that the intertwining character of the orbits remains intact even when projected on a two dimensional manifold. This makes the analysis easier. The template or knot holder actually summarizes the topological invariants and is usually associated with a template matrix and interwinding matrix. Template matrix is actually made up of linking and self linking numbers of different period one orbits and the interwinding matrix tells how these period ones are linked with one another. In the literature, depending on the form of the matrix and the data set the system is called non horseshoe.

People have already tried to apply these type of considerations for higher dimensional systems by means of suitable projections. In this connection the work of Bekkei [Bekkei, 2000] is worth mentioning.

Topological analysis procedure consists of the following successive steps.

(a) Closed Algorithm: This algorithm is used to locate the segment in the chaotic time series which can be used as surrogate for unstable periodic orbits which actually populate the attractor.

(b) Topological constant: The topological organization of the unstable periodic orbits extracted from the time series is determined by calculating linking numbers of all pairs of periodic orbits so extracted and their self linking numbers.

(c) Template identification: The form of the template or knot holder of all UPOs is obtained on the basis of linking.

(d) Template Verification: Once template have been tentatively identified it can be used to find the linking number and self-linking number. later they can be compared with those obtained from the surrogate orbit. if they match the template is alright.

## 5.4. Application to Circuit Data

### 5.4.1. *Search for Close Return*

Periodic orbit for a flow are reconstructed from directly sampled time series data $x(i)$ using a straight forward procedure. The time series data are scanned for close return (strong recurrence properties):

$$d[x(i) - x(i+n)] < \epsilon$$

Typically number of sampled steps $n$ between close return is an integer multiple of some smallest number, which can be associated with the fundamental period of periodically driven dynamical system. Segments of the period $k(= n/n_0)$ are compared, and those remain closed throughout the period are associated with the same unstable periodic orbit of period $k$. This unstable orbit is estimated by choosing the orbit with best recurrence properties

We now discuss how the above steps can be implemented for the data collected from the circuit given in Fig. 5.1. Such close return segment can be located in the original data as follows. If $x(i)$ (i=1,2,... N, N is the length of the data) and $x(i+p)$ are the coordinates of two points which are neighbor of some appropriate phase space; then $x(i+1)$ and $x(i+1+p)$ will have approximately equal values, as well $x(i+k)$ and $x(i+k+p)$ for some sequence of values of k=1,2, ..., where $p$ is the period of a nearby UPO, measured in unit by sampling time. Such segment can be recognized by collecting points with the property $|x(i) - x(i+p)| < \epsilon$, where $\epsilon$ is very small. In our case we used $\epsilon = 0.005$. The number of close return is plotted against $p$ in Fig. 5.4. Here this is evident that the minimum number of

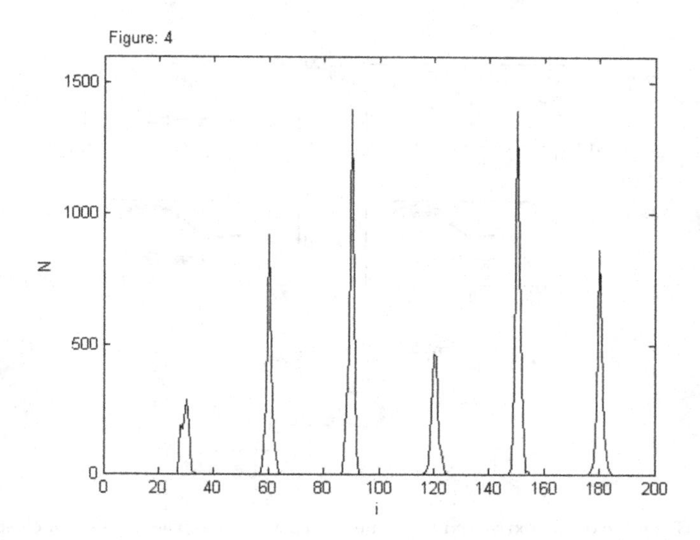


Fig. 5.4. Histogram plot of the close return obtained from the data at $a = 25.0, b = 8.1$ and $c = 45.0$.

time step (measured in unit of sampling time) at which maximum number of close return occurs is 30. So the period of period-1 orbit is 30 i.e. $n_0$. Other periods are its integer multiple. The details of the procedure can be found in Lathrop and Kostelich [Lathrop *et al.*, 1989]. Surrogate UPOs extracted through the above mentioned method are shown in Fig. 5.5.

It may be mentioned that the topological approach is based on the organizing the unstable periodic orbits whose linking properties severely constrain the structure of the strange attractor. A quantitative topological characterization of low dimensional chaotic sets requires a good symbolic encoding of the trajectories which is given by first return map build on the Poincaré section defined as follows:

$$P_x = \{(x, y, z) \in R^3 : \frac{dx}{dt} = 0\} \tag{5.4}$$

A mask of the attractor which may be viewed as the knot-holder of the chaotic system is build after many visual investigations in the tri-dimensional state space. This mask is related to the stretched and folded band on which asymptotic trajectories evolve in the state space. Such an approach can be used whenever the vector field is strongly dissipative and if the Lyapunov dimension is less then three. This method was first introduced by Birman and Williams on the Lorenz system. Once the knot holder is extracted, the topology is synthesized onto a template, which is

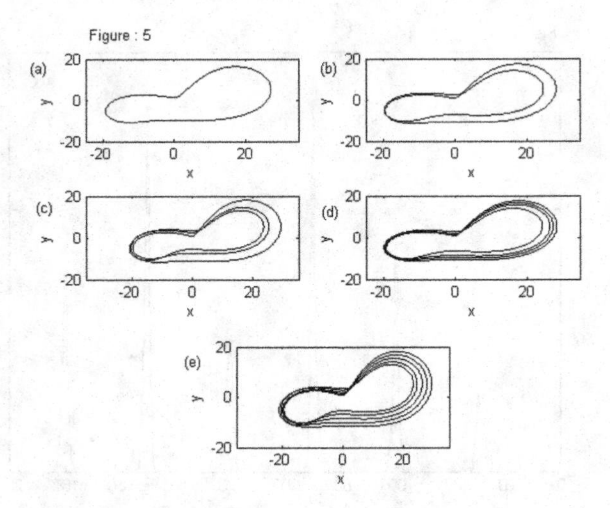


Fig. 5.5. Periodic orbits extracted from the data set through the method of close return. (a) period-1, (b) period-2, (c) period-3, (d) period-4, (e) period-5.

described by the linking matrix $M_{ij}$, in which $i$ and $j$ run from 0 to maximum symbolic name $n$ (for example 0 to 1 in our case).

In the last few decades, several research workers have studied the characteristics of a unimodal map under the variation of the control parameter. Such map is the actual output of the Poincaré section defined by Eq. (5.3a). The associated symbolic dynamics of the resultant unimodal map is useful to describe the creation of periodic orbits.

A unimodal map is actually divided into two parts increasing and decreasing half. These two are separated by critical point $C$ (which is the maxima in our case). The increasing and decreasing branches are labeled by 0 and 1 respectively. Then each point $x_n$ of the invariant set of the map possess a code $K(x_n) = \sigma_n$, defined by:

$$\sigma_n = \begin{cases} 0 & \text{if } x_n < C \\ 1 & \text{if } x_n > C \end{cases}$$

Consequently, a trajectory starting from $x$ with $k(x) = \sigma_1$ may be encoded by a string of successive codes reading as

$$\mathbf{S} = \sigma_1 \sigma_2 \ldots \sigma_i \ldots$$

In a period-$p$ orbit, a substring $\bar{\mathbf{S}}$ contains $p$ codes and reads as

$$\bar{\mathbf{S}} = \overline{\sigma_1 \sigma_2 \ldots \sigma_p}$$

with $\sigma_1 = \sigma_p$.

A periodic orbit may then be encoded by symbolic sequence (**W**) which is given by a sub-string $\bar{\mathbf{S}}$.

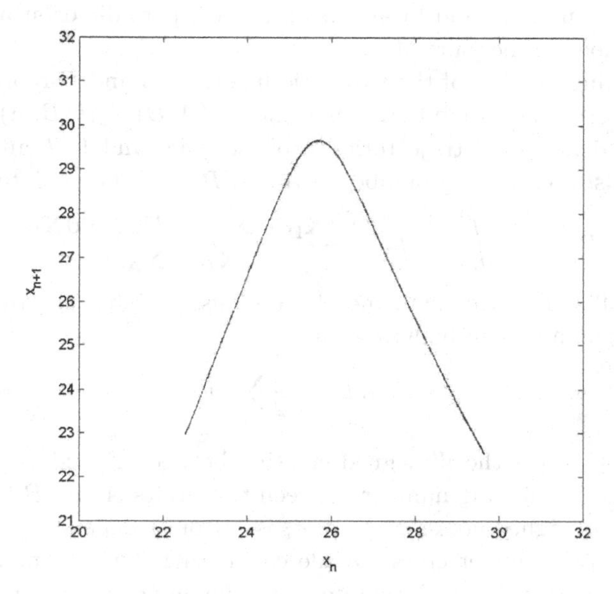


Fig. 5.6. Poincaré section obtained at $a = 25.0, b = 8.1$ and $c = 45.0$. Here the critical point i.e. maxima is at $x_c = 25.7$.

In this case, we consider the first return plot for the successive maxima and it is shown in Fig. 5.6. We define the maxima of such map as ($x_c = 25.7$). The increasing branch to the left of $x_c$ is defined with $'0'$ and decreasing branch is defined with $'1'$ which automatically preserve their parity. The increasing branch has even parity so it is denoted by even number. All the periodic orbit are encoded with binary symbols following the prescription given in the reference [Hao, 1989]. We have shown in Table.5.1, the state of 31 such periodic orbits extracted with the help of close return and encoded with binary symbolic dynamics.

### 5.4.2. *Topological Constant*

Now, for template identification we recapitulate a theorem due to Birman and Williams which greatly facilities the diagnosis of the dynamics of the system, which exhibits chaos and has hyperbolic invariant sets. It states that on can project the periodic orbits of a hyperbolic strongly contracting flow into a two dimensional projection keeping all the topological property intact. Once the UPOs are extracted from the attractor at a particular parameter values, one can go on to calculate the topological invariants like: i) Self linking number and local torsion of each periodic orbit and ii) The linking number of the pairs of orbits.

The linking number of the two periodic orbits $A$ and $B$ represents how many times $A$ wind around $B$. Obviously, $lk(A,B) = lk(B,A)$. If $\mathbf{X_A(t)}$ and $\mathbf{X_B(t)}$ denotes the trajectories in phase space and $P_AT$ and $P_BT$ are their periods, then linking number of $A$ and $B$ is the Gauss integral:

$$lk(A,B) = \frac{1}{4\pi} \int_o^{P_AT} \int_0^{P_BT} \frac{(\mathbf{X_B} - \mathbf{X_A}) \cdot (\mathbf{dX_A} \wedge \mathbf{dX_B})}{\|\mathbf{X_B} - \mathbf{X_A}\|^3} \tag{5.5}$$

But it is difficult to calculate the above Gaussian integral. To our relief, this linking number can be written as:

$$lk(A,B) = \frac{1}{2} \sum_i \sigma_i \tag{5.6}$$

where $\sigma_i$ represents the $i^{th}$ signed crossing between $A$ and $B$. Thus, one can calculate the linking number between two orbits $A$ and $B$ by counting the number of signed crossing. Here $\sigma_i$ is +1 or -1 depending on whether it is over cross or under cross. While calculating the self linking number for a periodic orbit $A$, which represents the linking of a periodic orbit with itself, we have to modify the Eq. (5.5) as

$$Slk(A) = \sum_i \sigma_i \tag{5.7}$$

But this self linking number is not a topological constant in $R^3$ space. This is constant in $R^2 \times S$ space.

We have calculated linking number and self linking number of a 31 orbits extracted from the dynamical system at ($a = 25.0, b = 8.1, c = 45.0$) and they are shown in Table. 5.2. The diagonal of this table gives the self linking number of the corresponding orbits and off-diagonal elements give the linking number between respective orbits. An example calculation is shown in Fig. 5.7. In Fig. 5.7a we have five black dots and one blank dot. This dots represent the crossing points of period three ($\overline{100}$). Black dots are positive crossings and blank dot is negative crossing. Thus, the total positive crossing number is 4. Thus, the self linking number of the corresponding periodic orbit is 4 (from Eq. (5.6)). In Fig. 5.7b, we have shown the linking between above period three orbit and period one orbit ($\overline{1}$). The total positive crossing number between one ($\overline{1}$) and three ($\overline{100}$) orbit is 6. Hence, the linking number is 3 (from Eq. (5.5)). At last we come

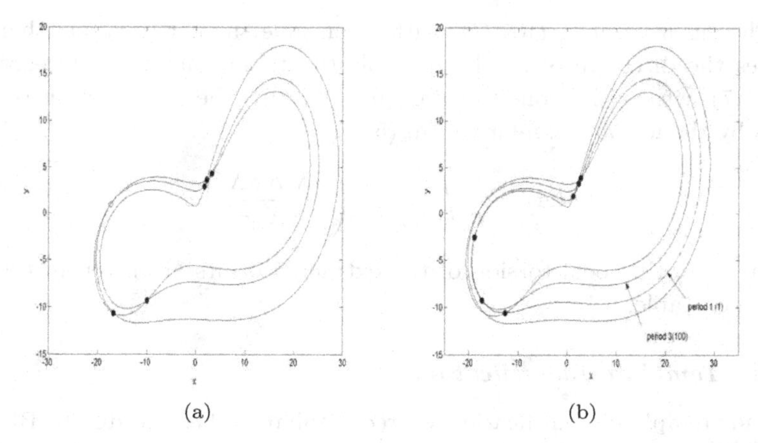


(a) (b)

Fig. 5.7. (a) Example of calculation of the self linking number of period-3 ($\overline{100}$). (b) Example of calculation of the linking number between previous period-3 orbit and period-1 ($\overline{1}$) orbit. Black dots show $+ve$ crossing and blank dot shows $-ve$ crossing..

to the part of calculating the local torsion. Local torsion measures which way the trajectories infinitely close to a periodic orbit wind around. As one follows the UPO over one period $P_A T$, the directions of the local stable ($W_l^s(A)$) and unstable ($W_l^u(A)$) manifold rotates by an integer number of half turns. This number is defined to be local torsion.

In numerical simulation $\Theta_l$ can be computed by using linearization of equation of motion around the periodic orbit. Given a set of ODEs

$$\frac{dX}{dt} = f(X,t) \tag{5.8}$$

The linearized equation govern the time evaluation of infinitesimal perturbation $\delta X$ of trajectory $X$:

$$\frac{d(\delta X(t))}{dt} = J(X,t)\delta X(t) \tag{5.9}$$

where Jacobian matrix is given by

$$J_{ij}(X,t) = \frac{\partial f_i(X,t)}{\partial X_j} \tag{5.10}$$

Given a periodic orbit $X_A(t)$ of periodic $P_A T$, its Floquet matrix $M_A(t)$, which expresses the linear relation between $X(t+P_A T)$ and $X(t)$ is given by:

$$\delta X(t + P_A T) = M_A(t) \times \delta X(t) \tag{5.11}$$

can be computed by integrating equation-21 over one period of the orbit for a basis of initial condition.

The eigenvector $\xi_s(t)(\xi_u(t))$ with eigenvalue smaller (greater) than indicates the direction of the local stable (unstable) manifold. Integrating Eq. (5.7) with initial condition $(\delta X(0) = \xi_u(0))$, the local torsion is then given by the formula similar to Eq. (5.6):

$$\Theta_l(A) = \frac{1}{\pi} \int_0^{P_A T} n \cdot \left( \frac{\delta X \wedge \delta \dot{X}}{\|\delta X\|^2} \right) dt$$

In our case, the local torsion of the extracted orbits are given in the $5^{th}$ collum of Table. 5.1.

### 5.4.3. *Template Identification*

Now for template identification we recapitulate a theorem due to Birman and Williams which greatly facilities the diagnosis of the dynamics of the system, which exhibits chaos and has hyperbolic invariant sets. It states that on can project the periodic orbits of a hyperbolic strongly contracting flow into a two dimensional projection keeping all the topological property intact.

The attractor is the result of scratching and folding of branches contained in it. To find the topological signature one computes the various linking numbers as coated above.

The actual template is represented with the help of a sequence matrix 'T' (template matrix) and row matrix 'l' (layering matrix), as we have a two latter symbolic dynamics we need a two branch template. Hence the template matrix is $2 \times 2$. Here $T(i,i)(i = 1,2)$ are the local torsion of the respective branches and $T(i,j)(i \neq j)$ are the two times the linking number between $i$ and $j$ branches.

Now, in our case for a given symbolic name and its (self) linking number, form Table. 5.1 and Table. 5.2 we have:

$$slk(01) = t_{01} + l_{01} = 1$$

$$lk(1,01) = \frac{1}{2}t_{01} + \frac{1}{2}t_{11} + \frac{\pi t_{11}}{2} l_{01} = 2$$

$$lk(1,0111) = \frac{1}{2}t_{01} + \frac{3}{2}t_{11} + \frac{\pi t_{11}}{2} l_{01} = 4$$

where $\pi(n) = 0$ or 1 if n is even or odd. From these equations we have the solutions

$$t_{01} = 2, t_{00} = 1, t_{11} = 2 \quad \text{and} \quad l_{00} = 0 \quad l_{01} = -1$$

Table 5.1. Number of UPOs of lowest periods of the attractor at $a = 25.0$, $b = 8.1$ and $c = 45.0$. The coordinate of outermost point of a periodic orbit are given. The $5^{th}$ collum shows the torsion of the periodic orbits and $6^{th}$ collum shows symbolic representation.

| Period | $x$ | $y$ | $z$ | Torsion | Symbol |
|---|---|---|---|---|---|
| 1 | 27.22653961 | 5.82991600 | 9.97722244 | 2 | 1 |
| 2 | 28.41831017 | 4.86866093 | 10.27653027 | 3 | 10 |
| 3 | 29.14678955 | 4.95937300 | 10.42918015 | 4 | 100 |
| 3 | 28.86454964 | 5.98923016 | 10.32627964 | 5 | 101 |
| 4 | 24.39430046 | 5.78171015 | 9.36116791 | 7 | 1011 |
| 4 | 22.85643005 | 5.02072716 | 9.06855679 | 6 | 1001 |
| 4 | 23.52930069 | 5.12330723 | 9.20887566 | 5 | 1000 |
| 5 | 24.48052979 | 5.42578602 | 9.39905834 | 8 | 10011 |
| 5 | 28.32622910 | 4.90867901 | 10.25510025 | 8 | 10110 |
| 5 | 24.19331932 | 5.03287601 | 9.35718346 | 9 | 10111 |
| 5 | 23.22990036 | 5.62540388 | 9.11529827 | 7 | 10010 |
| 6 | 24.08344078 | 5.51888895 | 9.30766201 | 9 | 100110 |
| 6 | 26.23648071 | 5.10144520 | 9.79607010 | 9 | 100101 |
| 6 | 24.60808945 | 5.73530817 | 9.41054058 | 8 | 100010 |
| 6 | 23.02000999 | 5.21304989 | 9.09308338 | 10 | 100111 |
| 6 | 22.61001968 | 5.43769121 | 8.99111462 | 9 | 100011 |
| 6 | 27.48838997 | 5.05273008 | 10.06863976 | 10 | 101110 |
| 7 | 22.99939919 | 5.36159706 | 9.08080482 | 10 | 1000110 |
| 7 | 22.53750038 | 5.22127485 | 8.98803711 | 11 | 1000111 |
| 7 | 23.08420944 | 5.39799595 | 9.09652519 | 12 | 1001111 |
| 7 | 23.10082054 | 5.30026388 | 9.10568619 | 11 | 1001110 |
| 7 | 23.49983978 | 5.65282297 | 9.17268753 | 10 | 1001100 |
| 7 | 23.28076935 | 5.13868999 | 9.15378189 | 10 | 1001010 |
| 7 | 28.84127045 | 5.15495396 | 10.35569954 | 12 | 1011011 |
| 8 | 23.28663063 | 5.40131903 | 9.14086533 | 11 | 10001010 |
| 8 | 24.42577934 | 5.01268578 | 9.40871906 | 13 | 10011101 |
| 8 | 24.05154037 | 5.17466593 | 9.31896210 | 13 | 10110110 |
| 8 | 24.07251930 | 5.35114002 | 9.31408310 | 12 | 10010110 |
| 8 | 23.03689957 | 5.77074718 | 9.06473064 | 13 | 10011110 |
| 8 | 29.63946915 | 5.51650810 | 10.51294994 | 12 | 10001110 |
| 8 | 22.56929016 | 5.06350899 | 9.00404739 | 13 | 10001111 |

Hence, the template can be written as

$$\begin{pmatrix} 1 & 2 \\ 2 & 2 \end{pmatrix}$$

$$\begin{pmatrix} 0 & -1 \end{pmatrix}$$

The two dimensional flow described by the above matrix is shown in Fig. 5.8. This kind of spiral template has rarely been seen in the literature. The arrow denotes the direction of the flow. On the left side the twisted part is

Table 5.2. The linking number and self linking number of UPOs of lowest period (at $a = 25.0$, $b = 8.1$ and $c = 45.0$) are given. UPOs are ordered in the same order as those of Table. 5.1.

| Period | 1 | 2 | 3 | 3 | 4 | 4 | 4 | 4 | 5 | 5 | 5 | 5 | 6 | 6 | 6 | 6 | 6 |
|---|---|---|---|---|---|---|---|---|---|---|---|---|---|---|---|---|---|
| 1 | 0 | 2 | 3 | 3 | 4 | 4 | 4 | 5 | 5 | 5 | 5 | 6 | 6 | 6 | 6 | 6 | 6 |
| 2 | 2 | 1 | 4 | 4 | 6 | 6 | 5 | 8 | 7 | 8 | 6 | 8 | 8 | 7 | 10 | 9 | 9 |
| 3 | 3 | 4 | 4 | 6 | 9 | 9 | 8 | 12 | 10 | 12 | 10 | 12 | 12 | 12 | 15 | 13 | 13 |
| 3 | 3 | 4 | 6 | 4 | 9 | 9 | 8 | 12 | 11 | 12 | 10 | 13 | 13 | 12 | 15 | 14 | 14 |
| 4 | 4 | 6 | 9 | 9 | 9 | 12 | 12 | 16 | 15 | 16 | 15 | 18 | 18 | 18 | 20 | 19 | 19 |
| 4 | 4 | 6 | 9 | 9 | 12 | 9 | 12 | 16 | 15 | 16 | 15 | 18 | 18 | 18 | 20 | 18 | 18 |
| 4 | 4 | 5 | 8 | 8 | 12 | 12 | 7 | 16 | 13 | 16 | 12 | 16 | 16 | 14 | 20 | 17 | 17 |
| 5 | 5 | 8 | 12 | 12 | 16 | 16 | 16 | 16 | 20 | 20 | 20 | 24 | 24 | 24 | 25 | 24 | 24 |
| 5 | 5 | 7 | 10 | 11 | 15 | 15 | 13 | 20 | 14 | 20 | 17 | 21 | 21 | 20 | 25 | 23 | 23 |
| 5 | 5 | 8 | 12 | 12 | 16 | 16 | 16 | 20 | 20 | 16 | 20 | 24 | 24 | 24 | 25 | 24 | 24 |
| 5 | 5 | 6 | 10 | 10 | 15 | 15 | 12 | 20 | 17 | 20 | 12 | 20 | 20 | 18 | 25 | 22 | 22 |
| 6 | 6 | 8 | 12 | 13 | 18 | 18 | 16 | 24 | 21 | 24 | 20 | 21 | 25 | 24 | 30 | 27 | 27 |
| 6 | 6 | 8 | 12 | 13 | 18 | 18 | 16 | 24 | 21 | 24 | 20 | 25 | 21 | 24 | 30 | 27 | 27 |
| 6 | 6 | 7 | 12 | 12 | 18 | 18 | 14 | 24 | 20 | 24 | 18 | 24 | 24 | 17 | 30 | 26 | 26 |
| 6 | 6 | 10 | 15 | 15 | 20 | 20 | 20 | 25 | 25 | 25 | 25 | 30 | 30 | 30 | 25 | 30 | 30 |
| 6 | 6 | 9 | 13 | 14 | 19 | 18 | 17 | 24 | 23 | 24 | 22 | 27 | 27 | 26 | 30 | 23 | 28 |
| 6 | 6 | 9 | 13 | 14 | 19 | 18 | 17 | 24 | 23 | 24 | 22 | 27 | 27 | 26 | 30 | 28 | 23 |

encapsulate the template and layering matrix. Rest of the part is preflow. There are two distinct branches '0' and '1'. The '0' branch is twisted about itself once. This represents the local torsion of branch '0'($t_{00} = 1$). Similarly the '1' branch is twisted about itself twice ($t_{11} = 1$). Now the two branches are linked with each other twice i.e. $t_{01} = 2$. Now come the bottom part. here strip '1' is in the background of '0' as $l_{01} = -1$ and $l_{00} = 0$.

### 5.4.4. *Template Verification*

The template matrix and the layering matrix obtained above were used to find the linking number and self-linking numbers of the different orbits and they are matched with values in Table. 5.2. The matching of these values verifies the template.

## 5.5. Conclusion and Discussion

In our above analysis we have constructed a new nonlinear system based on an electronic circuit made of usual op-amp, resistant, multiplexer capacitance etc. The data generated are collected and analyzed from the view point of the global chaotic scenario. We have extracted the UPOs and other symbolic sequence, which are latter used to compute self linking, cross link-

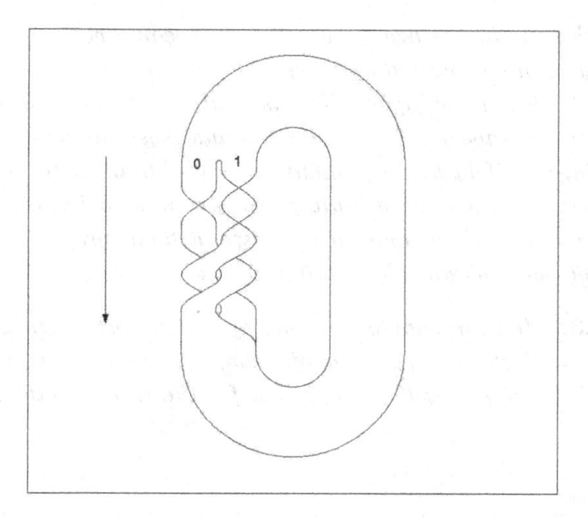


Fig. 5.8. The two dimensional figure of the template of the attractor obtained at $a = 25.0, b = 8.1$ and $c = 45.0$.

ing and torsion. The template matrix is then calculated and constructed. The form of template matrix and following diagrammatic figure suggests that we are dealing with non-horseshoe template.

Lastly, it can be ascertained that one of the few important problems in chaos theory are the identification of system with non-horseshoe chaos, its topological characterization, and subsequent identification of electronic circuit which will simulate such nonlinear system. In the present paper we have touched up an all these three aspects where the analysis is based on our new nonlinear system which has got higher order nonlinearity of our system with the Lorenz case can be seen from the projections of the attractors as observed on the oscilloscope. But the detailed topological analysis clearly exhibits a different and distinct structure.

We may conclude by noting some present problems in chaos research which need immediate attention:

**Problem 5.1.** *Circuit realization of bifurcation and Poincaré map, which are the basic of every instability study, are still not a very well studied matters.*

**Problem 5.2.** *Though above mentioned topological method gives very good result for any systems whose York dimension is less than three, it falls flat for higher dimensional systems. This is because of the fact that the links* *upon which this method is heavily based upon become very simple ('unknotted') for dimensions greater than three and so a global constant like linking number is still absent for higher dimensional system. People have tried to apply the above method for higher dimensional systems by projecting them on the inertial manifold whose dimension is less than three. But the applicability of that method is very limited. So we need to improve the method for higher dimensional systems to understand them properly. For that we must find a global constant like linking number.*

**Problem 5.3.** *Mention should be made of systems which involve delay. A proper topological characterization of delay system is still missing. So a large class of systems is still out of reach for the above method.*

## References


1. Bekki, N. [2000] "Torus knot in a dissipative fifth-order system," *J. Phys. Soc. Jpn* **69**, 295–298.
2. Birman, J. S. & Williams,R. F. [1983] "Knotted periodic orbits in dynamical systems. I. Lorenz's equations," *Topology* **22**, 47–82.
3. Boulant, G., Lefranc, M., Bielawski, S. & Derozier, D. [1997] "Experimental observation of a chaotic attractor with a reverse horseshoe topological structure," *Phys. Rev. E* **55**, R3801–R3804.
4. Boulant, G., Lefranc, M., Bielawski, S. & Derozier, D. [1997] "Horseshoe templates with global torsion in a driven laser," *Phys. Rev. E* **55**, 5082–5091.
5. Boulant, G., Lefranc, M., Bielawski, S. & Derozier, D. [1998] "A nonhorseshoe template in a chaotic laser model," *Int. J. Bifurcation & Chaos* **8**, 965–975.
6. Cuomo, K. A. & Oppenheim, A. V. [1993] "Circuit implementation of synchronized chaos with applications to communications," *Phys. Rev. Lett* **71**, 65–68.
7. Freire, E., Rodríguez-Luis, A. J., Gamero, E. & Ponce, E. [1993] "A case study for homoclinic chaos in an autonomous electronic circuit," *Phys. D* **62**, 230–253.
8. Gilmore, R. & McCallum, J. W. L. [1995] "Superstructure in the bifurcation diagram of the Duffing oscillator," *Phys. Rev. E* **51**, 935–956.
9. Gilmore, R. & Lefranc, M. [2001] *The Topology of Chaos*, 2th edn, Wiley, New York.
10. Hao, B. -L. [1989] *Elementary Symbolic Dynamics and Chaos in Dissipative System*, World Scientific, Singapore.
11. Kiers, K., Klein, T., Kolb, J., Price, S. & Sprott, J. C. [2004] "Chaos in a nonlinear analog computer," *Inter. J. Bifurcation & Chaos* **14**, 2867–2873.
12. Matsumoto, T., [1984] "A chaotic attractor from Chua's circuit," *IEEE Trans.Circuits & Systems* **31**, 1055–1058.

13. Mindlin, G. B., Hou, X. -J, Solari, H. G. and Gilmore, R. & Natiello, M. A. [1990] "Classification of strange attractors by integers," *Phys. Rev. E* **64**, 2350–2353.
14. Lathrop, D. P. & Kostelich, E. J. [1989] "Characterization of an experimental strange attractor by periodic orbits," *Phys. Rev. A* **40**, 4028–4031.
15. Lefranc, M. & Glorieux, P. [1993] "Topological analysis of chaotic signals from a $CO_2$ laser with modulated losses," *Int. J. Bifuration & Chaos* **3**, 643–650.
16. Letellier, C., Dutertre, P., Gouesbet, G. [1994] "Characterization of the Lorenz system, taking into account the equivariance of the vector field," *Phys. Rev. E* **49**, 3492–3495.
17. Letellier, C., Dutertre, P. & Maheu, B. [1995] "Unstable periodic orbits and templates of the Rössler system: Toward a systematic topological characterization," *Chaos* **5**, 271–282.
18. Mindlin, G. B., Solari, H. G., Natiello, M. A., Gilmore, R. & Hou, X. -J. [1991] "Topological analysis of chaotic time series data from the Belousov-Zhabotinskii reaction," *J. Nonlinear Science* **1**, 147–173.
19. Mindlin, G. B. & Gilmore, R. [1992] "Topological analysis and synthesis of chaotic time series," *Phys. D* **58**, 229–242.
20. Ray, A., Ghosh, D. & Roychowdhury, A. [2009] "Topological study of multiple coexisting attractors in a nonlinear system," *J. Phy. A* **42**, 385102–385117.
21. Solari, H. G. & Gilmore, R. [1988] "Relative rotation rates for driven dynamical systems," *Phys. Rev. A* **22**, 3096–3109.

# Chapter 6

# Instability of Solutions of Fourth and Fifth Order Delay Differential Equations

Cemil Tunç

*Department of Mathematics, Yüzüncü Yıl University*
*Faculty of Arts and Sciences, 65080, Van, Turkey*
*Email: cemtunc@yahoo.com*

The main purpose of this paper is to introduce some open problems related to some nonlinear differential equations of fourth and fifth order with constant delay.

**Keywords**: Instability, Lyapunov functional, delay differential equation, fourth order, fifth order.

## Contents



## 6.1. Introduction

It is well known that the investigations of qualitative behaviors of solutions of higher order nonlinear differential equations are very important problems in the theory and applications of differential equations. See, in particular the papers of Linz [2008], Chlouverakis & Sprott [2006] and the references cited in these papers.

With respect to our observation from literature, in the last three decades, many authors give much attention to investigate instability of solutions of nonlinear ordinary differential equations of fourth and fifth order without delay of the form:

$$x^{(4)}(t) + A_1\dddot{x}(t) + A_2\ddot{x}(t) + A_3\dot{x}(t) + A_4x(t) = 0$$

and

$$x^{(5)}(t) + B_1 x^{(4)}(t) + B_2 \dddot{x}(t) + B_3 \ddot{x}(t) + B_4 \dot{x}(t) + B_5 x(t) = 0$$

where $x \in \Re$, $t \in \Re_+$, $\Re_+ = [0, \infty)$, $A_1$ , $A_2$, $A_3$, $A_4$, $B_1$, $B_2$, $B_3$, $B_4$ and $B_5$ are not necessarily constants. For a comprehensive treatment of the subject we refer the reader to the papers of Ezeilo [1978a, 1978b, 1979a, 1979b, 2000], Skrapek [1980], Tiryaki [1988], Tunç [2004, 2005, 2006, 2008], Tunç and Erdogan [2007], Tunç and Karta [2008], Tunç and Şevli [2005], C. Tunç and E. Tunç [2004] and the references cited in these papers for some works performed on the topic, which include some nonlinear differential equations of fourth and fifth order without delay.

It should be also noted that throughout all of these papers, based on Krasovskii's properties (see Krasovskii [1955]), Lyapunov's [1966] second (or direct) method has been used as a basic tool to prove the main results. That is to say, if solutions of any differential equation under consideration are known in closed form, we can determine the instability properties of the system or the solutions of differential equation, appealing directly the definition of instability. However, in general, it is not always possible to find the solution of all linear or nonlinear differential equations. Moreover, finding solutions can be even more difficult for delay differential equations rather than ordinary differential equations. Therefore, it is crucial to obtain information on the instability of solutions of differential equations while we have no analytical expression for solutions. By Lyapunov's [1966] second (or direct) method, one can determine the instability of solutions by constructing a suitable auxiliary function or functional, without solving a differential equation without and with delay under consideration. That is, the method yields stability and instability information directly, i.e., without solving the differential equation. But, construction of appropriate Lyapunov functions and functionals for higher order ordinary or delay differential equations remain as a general problem in the literature. In the following theorems, we give basic idea of the method about the instability of solutions of ordinary and delay differential equations.

The following theorem, due to the Russian mathematician N. G. Cetaev's (see LaSalle & Lefschetz [1961]).

**Theorem 6.1.** ***Instability Theorem of Cetaev's*** *Let $\Omega$ be a neighborhood of the origin. Let there be given a function $V(x)$ and region $\Omega_1$ in $\Omega$ with the following properties:*

*(i) $V(x)$ has continuous first partial derivatives in $\Omega_1$.*

*(ii)* $V(x)$ *and* $\dot{V}(x)$ *are positive in* $\Omega_1$.
*(iii) At the boundary points of* $\Omega_1$ *inside* $\Omega$, $V(x) = 0$.
*(iv) The origin is a boundary point of* $\Omega_1$.
*Under these conditions the origin is an unstable.*

Let $r \geq 0$ be given, and let $C = (C[-r, 0],\ \Re^n)$ with

$$\|\phi\| = \max_{-r \leq s \leq 0} |\phi(s)|\,,\ \phi \in C.$$

For $H > 0$ define $C_H \subset C$ by

$$C_H = \{\phi \in C : \|\phi\| < H\}.$$

If $x : [-r,\ a] \to \Re^n$ is continuous, $0 < A \leq \infty$, then, for each $t$ in $[0, A)$, $x_t$ in $C$ is defined by

$$x_t(s) = x(t+s), -r \leq s \leq 0, t \geq 0.$$

Let $G$ be an open subset of $C$ and consider the general autonomous delay differential system with finite delay

$$\dot{x} = F(x_t), x_t = x(t+\theta), -r \leq \theta \leq 0, t \geq 0$$

where $F : G \to \Re^n$ is a continuous and maps closed and bounded sets into bounded sets. It follows from these conditions on $F$ that each initial value problem

$$\dot{x} = F(x_t), x_0 = \phi \in G$$

has a unique solution defined on some interval $[0, A)$, $0 < A \leq \infty$. This solution will be denoted by $x(\phi)(.)$ so that $x_0(\phi) = \phi$.

**Definition 6.1.** The zero solution, $x = 0$, of $\dot{x} = F(x_t)$ is stable if for each $\varepsilon > 0$ there exists $\delta = \delta(\varepsilon) > 0$ such that $\|\phi\| < \delta$ implies that $|x(\phi)(t)| < \varepsilon$ for all $t \geq 0$. The zero solution is said to be unstable if it is not stable.

**Theorem 6.2.** *Suppose there exists a Lyapunov function* $V : G \to \Re_+$ *such that* $V(0) = 0$ *and such that* $V(x) > 0$ *if* $x \neq 0$. *If either*
*(i)* $\dot{V}(\phi) > 0$ *for all* $\phi$ *in* $G$ *for which*

$$V[\phi(0)] = \max_{-s \leq t \leq 0} V[\phi(s)] > 0$$

*or*

*(ii)* $\dot{V}(\phi) > 0$ *for all* $\phi$ *in* $G$ *for which*

$$V[\phi(0)] = \min_{-s \leq t \leq 0} V[\phi(s)] > 0$$

*then* $x = 0$ *of* $\dot{x} = F(x_t)$ *is unstable (see [Haddock & Zhao, 2006]).*

It should be noted that the theory of Lyapunov functions and functionals is a global approach toward determining asymptotic behaviors of solutions. It is worth mentioning that constructions of that functions and functional remains as a general problem in the literature. This theory became an important part of both mathematics and theoretical mechanics in twentieth century

However, to the best of our knowledge, we could not see any instability results in the literature for fourth and fifth order delay differential equations of the forms:

$$x^{(4)}(t) + A_1\dddot{x}(t-r) + A_2\ddot{x}(t-r) + A_3\dot{x}(t-r) + A_4x(t-r) = 0$$

and

$$x^{(5)}(t)+B_1x^{(4)}(t-r)+B_2\dddot{x}(t-r)+B_3\ddot{x}(t-r)+B_4\dot{x}(t-r)+B_5x(t-r) = 0$$

where $x \in \Re$, $t \in \Re_+$, $\Re_+ = [0, \infty)$, $r$ is a positive constant fixed delay, $A_1$, $A_2$, $A_3$, $A_4$, $B_1$, $B_2$, $B_3$, $B_4$ and $B_5$ are not necessarily constants.

Our aim here is to introduce some open problems related to instability of the trivial solution of delay differential equations in the above forms. The motivation to write these problems for the delay differential equations in the above forms comes from the foregoing papers done for ordinary differential equations.

## 6.2. Open Problems

In 1978, Ezeilo [1978a] discussed instability of the trivial solution of the fourth-order scalar differential equation without delay:

$$x^{(4)} + a_1\dddot{x} + a_2\ddot{x} + a_3\dot{x} + f(x) = 0 \tag{6.1}$$

The system equivalent to Eq. (6.1) is given by:

$$\begin{aligned}\dot{x} &= y\\ \dot{y} &= z\\ \dot{z} &= w\\ \dot{w} &= -\,a_1w - \; a_2z - \; a_3z - f(x)\end{aligned}$$

The author established the following result for instability of the trivial solution $x = 0$ of Eq. (6.1). In [Ezeilo, 1978a], Ezeilo proved the following theorem:

**Theorem 6.3.** *If $f(0) = 0$ and if there is a constant $a_4 > \frac{1}{4}a_2^2$ such that*

$$f'(x) \geq a_4 \quad \textit{for all } x$$

*then the trivial solution of Eq. (6.1) is unstable.*

**Remark 6.1.** In order to prove Theorem 6.3, it is sufficient (see [Krasovskii, 1955]) to show that there exists a continuous Lyapunov function $V(x, y, z, w)$ which satisfies the following Krasovskii properties:

$(K_1)$ In every neighborhood of $(0, 0, 0, 0)$, there exists a point $(\xi, \eta, \zeta, \mu)$ such that $V(\xi, \eta, \zeta, \mu) > 0$;

$(K_2)$ The time derivative $\frac{d}{dt}V(x, y, z, w)$ along solution paths of the above system is positive semi-definite; and

$(K_3)$ The only solution $(x, y, z, w) = (x(t), y(t), z(t), w(t))$ of the above system (6.1) which satisfies $\frac{d}{dt}V(x, y, z, w)$ is the trivial solution $(0, 0, 0, 0)$.

We now consider fourth order nonlinear delay differential equation

$$x^{(4)} + a_1\dddot{x} + a_2\ddot{x} + a_3\dot{x} + f(x(t-r)) = 0 \tag{6.2}$$

where $r$ is positive constant, a fixed delay, $a_1$, $a_2$ and $a_3$ are some constants; the dots in Eq. (6.2) denote differentiation with respect to $t$, $t \in \Re_+$, $\Re_+ = [0, \infty)$; $f$ is a continuous function on $\Re$ with $f(0) = 0$. Our first open problem is the following one:

**Problem 6.1.** *What are delay dependent or independent conditions for the trivial solution of Eq. (6.2) to be unstable, or when this equation has a chaotic and bounded orbit?*

Later, Ezeilo [1979b] discussed instability of the trivial solution of a more general ordinary differential equation of the form:

$$x^{(4)} + a_1\dddot{x} + h(x, \dot{x}, \ddot{x}, \dddot{x})\ddot{x} + g(x)\dot{x} + f(x) = 0,\ f(0) = 0 \tag{6.3}$$

The author established the following theorem:

**Theorem 6.4.** *If*

$$x^{-1}f(x) > \frac{1}{4}h^2(x, y, z, w) > 0 \quad \textit{for arbitrary } x(\neq 0),\ \ y,\ z,\ w$$

*then the trivial solution $x = 0$ of Eq. (6.3) is unstable for every $a_1$, and $g(x)$.*

We now consider fourth order nonlinear delay differential equation:

$$\begin{cases} x^{(4)} + a_1\dddot{x} + l(t)\ddot{x} + g(x(t-r))\dot{x} + f(x(t-r)) = 0 \\ \quad l(t) = h(x(t-r), \dot{x}(t-r), \ddot{x}(t-r), \dddot{x}(t-r)) \end{cases} \tag{6.4}$$

where $r$ is a fixed positive constant delay, $a_1$ is a constant; the dots in Eq. (6.4) denote differentiation with respect to $t$, $t \in \Re_+$, $\Re_+ = [0, \infty)$; $h$, $g$, $f$ are continuous functions on $\Re^4$, $\Re$, $\Re$ respectively with $f(0) = 0$.

Our second open problem is the following one:

**Problem 6.2.** *What are delay dependent or independent conditions which guarantee that the trivial solution of Eq. (6.4) is unstable, or when this equation has a chaotic and bounded orbit?*

Tiryaki [1988] interested in a fourth-order nonlinear ordinary differential equation of the form:

$$x^{(4)} + \psi(\ddot{x})\dddot{x} + \varphi(\dot{x})\ddot{x} + \theta(\dot{x}) + f(x) = 0 \tag{6.5}$$

Under the specified conditions imposed on the functions $\psi$, $\varphi$, $\theta$ and $f$, the author established the following sufficient conditions that guarantee instability of the trivial solution of Eq. (6.5):

**Theorem 6.5.** *Let* $\Phi(y) \equiv \int_0^y \varphi(\eta)d\eta$. *If*

$$f'(x) - \frac{1}{4}y^{-2}\Phi^2(y) > 0, \quad (y \neq 0)$$

*for an arbitrary* $x$, *then the trivial solution* $x = 0$ *of Eq. (6.5) is unstable for all arbitrary* $\psi$.

Instead of Eq. (6.5), we take in consideration the following fourth-order nonlinear delay differential equation:

$$\begin{cases} x^{(4)} + \psi(\ddot{x}(t-r))\dddot{x} + m\,(t) = 0 \\ m\,(t) = \varphi(\dot{x}(t-r)\ddot{x} + \theta(\dot{x}(t-r)) + f(x(t-r)) \end{cases} \tag{6.6}$$

where $r$ is a positive constant, fixed delay, $t \in \Re_+$, $\Re_+ = [0, \infty)$; $\psi$, $\varphi$, $\theta$, $f$ are all continuous functions defined on $\Re$ with $\theta(0) = f(0) = 0$.

Our third open problem is the following one:

**Problem 6.3.** *What are delay dependent or independent conditions which guarantee that the trivial solution of Eq. (6.6) is unstable, or when this equation has a chaotic and bounded orbit?*

Later, Ezeilo [2000] discussed a similar problem for fourth order ordinary differential equations:

$$x^{(4)} + \psi(\ddot{x})\dddot{x} + g(x, \dot{x}, \ddot{x}, \dddot{x})\ddot{x} + \theta(\dot{x}) + f(x) = 0 \tag{6.7}$$

and

$$x^{(4)} + a_1\dddot{x} + g(x, \dot{x}, \ddot{x}, \dddot{x})\ddot{x} + h(x)\dot{x} + f(x, \dot{x}, \ddot{x}, \dddot{x}) = 0 \qquad (6.8)$$

respectively, where $a_1$ is a constant. Ezeilo [2000] proved the following instability theorems:

**Theorem 6.6.** *Suppose there exists a constant $a_2$ such that*

$$a_2 > 0, \; g(x, y, z, w) \le a_2 \textit{ for all } x, y, z, w,$$

$$f'(x) > \frac{1}{4}a_2^2 \textit{ for all } x$$

*Then the trivial solution $x = 0$ of Eq. (6.7) is unstable.*

**Theorem 6.7.** *Suppose that*

$$g(0, y, z, w) = 0 \textit{ for arbitrary } y, z, w$$

*and*

$$x^{-1}f(x, y, z, w) > \frac{1}{4}g^2(x, y, z, w) \textit{ for arbitrary } x\ (\neq 0),\ y,\ z,\ w$$

*Then the trivial solution $x = 0$ of Eq. (6.8) is unstable for arbitrary $a_1$, and $h(x)$.*

Instead of Eq. (6.7) and (6.8), we take in consideration the following The system equivalent to Eq. (6.8)

$$\begin{cases} x^{(4)} + \psi(\ddot{x}(t-r))\dddot{x} + c(t)\,\ddot{x} + \theta(\dot{x}(t-r)) + f(x(t-r)) = 0 \\ \qquad c(t) = g(x(t-r), \dot{x}(t-r), \ddot{x}(t-r), \dddot{x}(t-r)) \end{cases} \qquad (6.9)$$

and

$$\begin{cases} x^{(4)} + a_1\dddot{x} + g_1(x(t-r), \dot{x}(t-r), \ddot{x}(t-r), \dddot{x}(t-r))\ddot{x} + b(t) = 0 \\ \quad b(t) = h(x(t-r))\dot{x} + \varphi(x(t-r), \dot{x}(t-r), \ddot{x}(t-r), \dddot{x}(t-r)) \end{cases} \qquad (6.10)$$

where $r$ is a fixed positive constant delay, $a_1$ is a constant; $t \in \Re_+$, $\Re_+ = [0, \infty)$; $\psi$, $g$, $\theta$, $f$, $g_1$, $h$, $\varphi$ are continuous functions in their respective arguments with $\theta(0) = f(0) = \varphi(0, \dot{x}, \ddot{x}, \dddot{x}) = 0$.

Our fourth and fifth open problems are given as follow:

**Problem 6.4.** *What are delay dependent or independent conditions which guarantee that the trivial solution of Eq. (6.9) is unstable, or when this equation has a chaotic and bounded orbit?*

**Problem 6.5.** *What are delay dependent or independent conditions which guarantee that the trivial solution of Eq. (6.10) is unstable, or when this equation has a chaotic and bounded orbit?*

Ezeilo [1978b] studied instability of the trivial solution $x = 0$ of the following ordinary differential equation of the fifth order:

$$x^{(5)} + a_1 x^{(4)} + a_2 \dddot{x} + a_3 \ddot{x} + a_4 \dot{x} + f(x) = 0, f(0) = 0 \tag{6.11}$$

where $a_1$, $a_2$, $a_3$, $a_4$ are some constants.
Ezeilo [1978b] introduced the following sufficient conditions for instability of the trivial solution of Eq. (6.11).

**Theorem 6.8.** *The trivial solution of Eq. (6.11) is unstable if*
*(i)* $a_1 > 0$, $f(0) = 0$, $f'(x) > \delta_5$ *for all* $x$, *where*

$$\delta_5 = \begin{cases} 0, & \text{if } a_3 \le 0, \\ a_3^2 \left|a_1^{-1}\right|, & \text{if } a_3 > 0 \end{cases}$$

*or if*
*(ii)* $a_1 < 0$, $f(0) = 0$, $f'(x) < -\delta_5'$ *for all* $x$, *where*

$$\delta_5' = \begin{cases} 0, & \text{if } a_3 \le 0, \\ a_3^2 \left|a_1^{-1}\right|, & \text{if } a_3 > 0 \end{cases}$$

We now consider the following non-linear delay differential equation:

$$x^{(5)} + a_1 x^{(4)} + a_2 \dddot{x} + a_3 \ddot{x} + a_4 \dot{x} + f(x(t-r)) = 0 \tag{6.12}$$

where $r > 0$, $a_1$, $a_2$, $a_3$ and $a_4$ are some constants; the dots in Eq. (6.12) denote differentiation with respect to $t$, $t \in \Re_+$, $\Re_+ = [0, \infty)$; $f$ is a continuous function on $\Re$ with $f(0) = 0$. Our sixth open problem is given as the following.

**Problem 6.6.** *What are delay dependent or independent conditions which guarantee that the trivial solution of Eq. (6.12) is unstable, or when this equation has a chaotic and bounded orbit?*

Later, Ezeilo [1978b] also investigated the following ordinary differential equations of fifth order:

$$x^{(5)} + a_1 x^{(4)} + a_2 \dddot{x} + h(\dot{x})\ddot{x} + g(x)\dot{x} + f(x) = 0 \tag{6.13}$$

and

$$x^{(5)} + \psi(\ddot{x})\dddot{x} + \phi(\ddot{x}) + \theta(\dot{x}) + f(x) = 0 \tag{6.14}$$

Ezilo [1978] proved the following theorems.

**Theorem 6.9.** *Let* $H(y) \equiv \int_0^y h(\eta)d\eta$. *Then trivial solution of Eq. (6.13) is unstable if*

*(i)*

$$a_1 > 0 \ \ \textit{and}\ f(x),\ H(y) \ \ \textit{satisfy}$$

$$xf(x) > 0, \quad (x \neq 0)$$

$$yH(y) \leq 0 \ \ \textit{for all}\ y$$

*or if (ii)*

$$a_1 < 0 \ \ \textit{and}\ f(x),\ H(y) \ \ \textit{satisfy}$$

$$xf(x) < 0, \quad (x \neq 0)$$

$$yH(y) \geq 0 \ \textit{for all}\ y$$

**Theorem 6.10.** *The trivial solution of Eq. (6.14) is unstable if (i)* $f'(x) > 0$ *and* $z\phi(z) \leq 0$ *for all* $x,\ z$
*or if*

*(ii)* $f'(x) < 0$ *and* $z\phi(z) \geq 0$ *for all* $x,\ z$

Consider the following non-linear delay differential equations of fifth order:

$$x^{(5)} + a_1 x^{(4)} + a_2 \dddot{x} + h(\dot{x}(t-r))\ddot{x} + g(x(t-r))\dot{x} + f(x(t-r)) = 0 \tag{6.15}$$

and

$$x^{(5)} + \psi(\ddot{x}(t-r))\dddot{x} + \phi(\ddot{x}(t-r)) + \theta(\dot{x}(t-r)) + f_1(x(t-r)) = 0 \tag{6.16}$$

where $r > 0$, $a_1$ and $a_2$ are some constants; the dots in Eq. (6.15) and (6.16) denote differentiation with respect to $t$, $t \in \Re_+$, $\Re_+ = [0, \infty)$; $h,\ g,\ f,\ \psi,\ \phi,\ \theta,\ f_1$ are continuous functions on $\Re$ with $f(0) = \phi(0) = \theta(0) = f_1(0) = 0$. Our seventh and eighth open problems are given as follow:

**Problem 6.7.** *What are delay dependent or independent conditions for the trivial solution of Eq. (6.15) to be unstable, or when this equation has a chaotic and bounded orbit?*

**Problem 6.8.** *What are delay dependent or independent conditions for the trivial solution of Eq. (6.16) to be unstable, or when this equation has a chaotic and bounded orbit?*

Later, in 1979, Ezeilo [Ezeilo, 1978] gave an instability result for the fifth order differential equation:

$$x^{(5)} + a_1 x^{(4)} + a_2 \dddot{x} + g(\dot{x})\ddot{x} + h(x, \dot{x}, \ddot{x}, \dddot{x}, x^{(4)}) + f(x) = 0 \tag{6.17}$$

by the following theorem:

**Theorem 6.11.** *If we have that*

$$f(0) = 0, f(x) \neq 0 \text{ for all } x \neq 0, yh(x, 0, z, w, u) = 0$$

*and there exists a constant* $a_4 > \frac{1}{4}a_2^2$ *such that*

$$yh(x, y, z, w, u) \geq a_4 y^2 (\text{ for} x,\ y, z, w, u)$$

*then the trivial solution* $x = 0$ *of Eq. (6.17) is unstable.*

We now introduce the following fifth order delay differential equation:

$$\begin{cases} x^{(5)} + a_1 x^{(4)} + a_2 \dddot{x} + n(t) + f(x(t-r)) = 0 \\ n(t) = g(\dot{x}(t-r))\ddot{x} + h(x(t-r), \dot{x}(t-r), \ddot{x}(t-r), \dddot{x}(t-r), x^{(4)}(t-r)) \end{cases} \tag{6.18}$$

where $r > 0$, $a_1$ and $a_2$ are some constants; the dots in Eq. (6.18) denote differentiation with respect to $t$, $t \in \Re_+$, $\Re_+ = [0, \infty)$; $g, h, f$ are continuous functions in their respective arguments with $h(x, 0, \ddot{x}, \dddot{x}, x^{(4)}) = f(0) = 0$. Our ninth open problem is given as the following:

**Problem 6.9.** *What are delay dependent or independent conditions which guarantee that the trivial solution of Eq. (6.18) is unstable, or when this equation has a chaotic and bounded orbit?*

Ezeilo [1979b] also considered the equation:

$$x^{(5)} + a_1 x^{(4)} + +a_2 \dddot{x} + \psi(x, \dot{x}, \ddot{x}, \dddot{x}, x^{(4)})\ddot{x} + \phi(x)\dot{x} + f(x) = 0 \tag{6.19}$$

and proved the following theorem:

**Theorem 6.12.** *If*

$$x^{-1}.f(x).sgna_1 > \frac{1}{4}\psi^2(x, y, z, w, u)\,|a_1|^{-1} > 0 \text{ for } x(\neq 0),\ \ y,\ z,\ w,u$$

*then trivial solution* $x = 0$ *of Eq. (6.19) is unstable for all arbitrary* $\phi(x)$.

Finally, we consider the following non-linear equation with a constant deviating argument, $r$ :

$$\begin{cases} x^{(5)} + a_1 x^{(4)} + +a_2 \dddot{x} + v\,(t) + \phi(x(t-r))\dot{x} + f(x(t-r)) = 0 \\ v\,(t) = \psi(x(t-r), \dot{x}(t-r), \ddot{x}(t-r), \dddot{x}\,(t-r), x^{(4)}(t-r))\ddot{x} \end{cases} \tag{6.20}$$

where $r > 0$, $a_1$ and $a_2$ are some constants; the dots in Eq. (6.20) denote differentiation with respect to $t$, $t \in \Re_+$, $\Re_+ = [0, \infty)$; $\psi$, $\phi$, $f$ are continuous functions in their respective arguments with $f(0) = 0$.

Our last, tenth, open problem is given as follow:

**Problem 6.10.** *What are delay dependent or independent conditions for the trivial solution of Eq. (6.20) to be unstable, or when this equation has a chaotic and bounded orbit?*

### 6.3. Conclusion

Some nonlinear differential equations of fourth and fifth order with constant delay are considered. Ten open problems with respect to these equations are introduced. By using the Lyapunov's second method, one can solve these problems (instability of equilibria).

### References


1. Chlouverakis, K. E. & Sprott, J. C. [2006] "Chaotic hyperjerk systems," *Chaos, Solitons & Fractals* **28**(3), 739–746.
2. Ezeilo, J. O. C. [1978] "An instability theorem for a certain fourth order differential equation," *Bull. London Math. Soc***10**(2), 184–185.
3. Ezeilo, J. O. C. [1978] "Instability theorems for certain fifth–order differential equations," *Math. Proc. Cambridge Philos. Soc* **84**(2), 343–350.
4. Ezeilo, J. O. C. [1979] "A further instability theorem for a certain fifth-order differential equation," *Math. Proc. Cambridge Philos. Soc* **86**(3), 491–493.
5. Ezeilo, J. O. C. [1979] "Extension of certain instability theorems for some fourth and fifth order differential equations," *Atti. Accad. Naz. Lincei. Rend. Cl. Sci. Fis. Mat. Natur* (**66**(4), 239–242.
6. Ezeilo, J. O. C. [2000] "Further instability theorems for some fourth order differential equations," *J. Nigerian Math. Soc***19**, 1–7.
7. Krasovskii, N. N. [1955] "On conditions of inversion of A. M. Lyapunov's theorems on instability for stationary systems of differential equations," *(Russian) Dokl. Akad. Nauk. SSSR (N.S)* **101**, 17–20.
8. Haddock, J. R. & Zhao, J. [2006] "Instability for functional differential equations," *Math. Nachr* **279**(13–14), 1491–1504.

9. LaSalle, J. & Lefschetz, S. [1961] *Stability by Liapunov's direct method with applications*, Mathematics in Science and Engineering **4** Academic Press, New York.
10. Linz, S. J. [2008] "On hyperjerky systems," *Chaos, Solitons & Fractals* **37**(3), 741–747.
11. Lyapunov, A. M. [1966] *Stability of Motion*, Academic Press, London.
12. Skrapek, W. A. [1980] "Instability results for fourth-order differential equations," *Proc. Roy. Soc. Edinburgh Sect. A* **85**(3–4), 247–250.
13. Tiryaki, A. [1988] "Extension of an instability theorem for a certain fourth order differential equation," *Bull. Inst. Math. Acad. Sinica* **16**(2), 163–165.
14. Tunc, C. [2004] "On the instability of solutions of certain nonlinear vector differential equations of fifth order," *Panamer. Math. J* **14**(4), 25–30.
15. Tunc, C. [2005] "An instability result for a certain non-autonomous vector differential equation of fifth order," *Panamer. Math. J* **15**(3), 51–58.
16. Tunc, C. [2006] "A further instability result for a certain vector differential equation of fourth order," *Int. J. Math. Game Theory Algebra* **15**(5), 489–495.
17. Tunc, C. [2008] "Further results on the instability of solutions of certain nonlinear vector differential equations of fifth order," *Appl. Math. Inf. Sci* **2**(1), 51–60.
18. Tunc, C. & Erdogan, F. [2007] "On the instability of solutions of certain non-autonomous vector differential equations of fifth order," *SUT J. Math* **43**(1), 35–48.
19. Tunc, C. & Karta, M. [2008] "A new instability result to nonlinear vector differential equations of fifth order," *Discrete Dyn. Nat. Soc* Art. ID 971534, 6pp.
20. Tunc, C. & Şevli, H. [2005] "On the instability of solutions of certain fifth order nonlinear differential equations," *Mem. Differential Equations Math. Phys* **35**, 147–156.
21. Tunc, C. & Tunc, E. [2004] "A result on the instability of solutions of certain non-autonomous vector differential equations of fourth order," *East–West J. Math* **6**(2), 153–160.

# Chapter 7

## Some Conjectures About the Synchronizability and the Topology of Networks

Acilina Caneco[1], J. Leonel Rocha[2], Clara Grácio[3] and Sara Fernandes[4]

[1]*Instituto Superior de Engenharia de Lisboa, Mathematics Unit DEETC and CIMA-UE, Rua Conselheiro Emidio Navarro, 1 1949-014 Lisboa, Portugal*
*E-mail: acilina@deetc.isel.ipl.pt*

[2]*Instituto Superior de Engenharia de Lisboa, Mathematics Unit DEQ, and CEAUL, Rua Conselheiro Emidio Navarro, 1 1949-014 Lisboa, Portugal*
*E-mail: jrocha@deq.isel.ipl.pt*

[3]*Department of Mathematics, Universidade de Évora and CIMA-UE Rua Romão Ramalho, 59, 7000-671 Évora, Portugal*
*E-mail: mgracio@uevora.pt*

[4]*Department of Mathematics, Universidade de Évora and CIMA-UE Rua Romão Ramalho, 59, 7000-671 Évora, Portugal*

We study the network synchronizability in terms of its local dynamics and in terms of its global dynamics. In previous work, we obtained some results about the synchronization interval, when we fix the graph underlying the network, i.e., the network connection topology and vary the dynamic in the nodes and about the effect of some graph parameters on the synchronization interval, supposing that the local dynamics in the nodes is fixed. We present numerical simulations for several network types and some conjectures suggested by these simulations. In particular, we are interested in the effect of the clustering coefficient, the conductance and the clustering performance, on the network synchronizability.

**Keywords**: Network, synchronization, chaos, topological entropy, clustering, conductance.

## Contents



## 7.1. Introduction

Complex dynamical networks are ubiquitous in the real world [Boccaletti *et al.*, 2006]. Typical examples of complex networks include the Internet, the World Wide Web, communication and transportation networks, power grids, biological neural networks, trading market trains, scientific citation networks, social relationship networks, among many others. The behavior of a complex dynamical network is a fundamental topic of considerable interest within science and technology communities. Although features of these networks have been studied in the past, it was only recently that massive amount of data are available and computer processing is possible to more easily analyze the behavior of these networks and verify the applicability of the proposed models.

A network with a complex topology is mathematically described by a graph [Bollobás, 1985]. Classical random graphs were studied by Paul Erdős and Alfréd Rényi in the late 1950's. A complex network typically refers to an ensemble of dynamical units with nontrivial topological features that do not occur in simple systems such as completely regular lattices or completely random structures. In 1998 Watts and Strogatz [Watts & Strogatz, 1998] proposed the new small-world model to describe many of real networks around us and in 1999 Barabási and Albert [Barabási & Albert, 1999] proposed the new scale-free model based on preferential attachment. These models reflect the natural and man-made networks more accurately than the classical random graph model. This preferential attachment characteristic leads to the formation of clusters, the nodes with more links have a greater probability of getting new ones.

Network synchronization is one of the basic motions in nature where many connected systems evolve in time, with completely different behavior, but after some time they adjust a given property of their motion to a common behavior ([Feng *et al.*, 2007; Li & Chen, 2003; Watts & Strogatz, 1998; Caneco *et al.*, 2008-2009]). As described in [Pikovsky *et al.*, 2001], synchronization is an adjustment of rhythms of oscillating objects due to their weak interaction. In fact, the synchrony of networks can well explain not only many natural phenomena [Strogatz, 2003], but also has many applications, such as secure communications, synchronous information exchange in the internet and the synchronous digital transfer in communications networks [Miliou *et al.*, 2007], [Kolumban *et al.*, 1997], [Stavrinides *et al.*, 2009]. The coupling of two or more oscillators may be bidirectional (undirected graphs), if each one influences the others, or unidirectional or master-slave (digraphs or directed graphs) if in each connection between two oscillators, only one of the oscillators influences the other. If the coupled systems undergo a chaotic behavior and they became synchronized, this is called a chaotic synchronization. Chaotic deterministic signals exhibit several intrinsic features, beneficial to secure communication systems, both analog and digital ones.

The analysis of synchronization phenomena of dynamical systems started in the $17^{th}$ century with the finding of Huygens that two very weakly coupled pendulum clocks become synchronized in phase. Since then, several problems concerning the synchronization have been investigated, especially to know for what values of the coupling parameter there is synchronization. In [Pecora & Carroll, 1990-1991], the master stability method was derived. Synchronization of networks occurs for values of coupling parameter belonging to a certain interval. The extremes of this interval depend not only on the local dynamical nodes, expressed by the maximum Lyapunov exponent of the individual chaotic nodes, but also on the network topology, expressed by spectrum of the Laplacian of the associated graph [Li & Chen, 2003].

The study of network synchronizability may be addressed in two approaches. One is fixing the connection topology and vary the local dynamics in the nodes and the other is consider the local dynamic fixed and vary the global dynamics, i.e., the structure of the connections.

In previous works [Caneco *et al.*, 2009-2010], we study the synchronization interval considering fixed the network connection topology, for different kinds of local dynamics. Supposing in the nodes, identical piecewise linear expanding maps, with different slopes in each subinterval, we obtained the synchronization interval in terms of the Lyapunov exponents of these maps. As a particular case, we derive the synchronization interval in terms of the topological entropy for piecewise linear maps with slope $\pm s$ everywhere and we proved that the synchronizability decreases if the local topological entropy increases. Considering identical chaotic symmetric bimodal maps in the nodes of the network, we express the synchronization interval in terms of one single critical point of the map.

In this work, we present some conjectures concerning the relation of some graph invariants with the spectrum of the Laplacian matrix. We perform experimental evaluations, that deepens the understanding of the effect of the conductance and the clustering coefficient on the network synchronizability. We obtain numerical evidence that, while there is no cluster formation, the network synchronizability becomes poorer if the clustering coefficient or the conductance decreases, but the relation between the synchronizability and the clustering coefficient reverses, when the cluster formation is apparent. To quantify this situation, we introduce the concept of clustering performance, which characterizes the clustering quality. We obtain numerical evidence that the amplitude of the synchronization interval decreases when the quality of the clustering increases.

The layout of this paper is as follows. In Sec. 7.2. a brief history of synchronization studies and related problems is presented. In Sec. 7.3 some applications of networks and graphs are given, showing the importance of synchronization. In Sec. 7.4., we present some notions and basic results on graphs and discrete dynamical systems. Sec. 7.5. is dedicated to the study of complete clustered networks. In Sec. 7.5.1. we define the clustering point as the point at which begins the formation of clusters in a network and we find a relation of this point with the number of intra-cluster edges, the number of inter-cluster edges and the order of the network. We suggest a formula to identify exactly this point, which determines the maximum number of edges that the graph may have, in order to be visible the formation of clusters. In Sec. 7.5.2. we study the importance of graph invariants, such as conductance, coefficient of clustering and performance of a cluster in the formation of a complete clustered network. It is possible to establish a classification of the clustering (its performance), the behavior of the conductance and the amplitude of the synchronization interval. In Sec. 7.5.3 a discussion about these problems is made. While in Sec. 7.5. it is considered the study of the effect of some graph parameters on the synchronization interval, supposing that the local dynamics in the nodes is fixed, in Sec. 7.6. we conjecture some results about the synchronization interval, when we fix the graph underlying the network, i.e., the network connection topology and vary the dynamic in the nodes. It is analyzed how the amplitude of the synchronization interval varies, if we consider a network with a fixed topology, having in each node an element on some lines of the tree of admissible trajectories of unimodal maps. Our goal is to relate the synchronization interval, with the topological entropy in each node of a network with identical chaotic nodes.

## 7.2. Related and Historical Problems About Network Synchronizability

To describe the complex evolution of a changing process, one often compare it with some other processes. One obtains the similarities and differences, for a time interval, between a dynamical system and another known dynamical system. Such similarity in a certain time interval is a kind of synchronization. The synchronization of two or more dynamical systems is a basis to understand an unknown dynamical system from one or more well-known dynamical systems. So, the concept of synchronization in dynamical systems is a universal concept for dynamical systems.

There are four principal classes of synchronization of two or more dynamical systems: (i) identical or complete synchronization, (ii) generalized synchronization, (iii) phase synchronization, and (iv) anticipated and lag synchronization and amplitude envelope synchronization. Once the two or more systems form a state of synchronization, such a state should be stable. See [Pikovsky *et al.*, 2001; Boccaletti, 2008].

In 1920s, the synchronization was stimulated by the development of electrical and radio wave propagations. After the Huygens investigation, Rayleigh, [Rayleigh, 1945], worked on the theory of sound which describes synchronization in acoustic. In the beginning, the investigation on synchronizations focus on the limit cycles in self-excited dynamical systems, resonance phenomena in multiple-degrees of freedom systems and, steady-state motion in forced vibration. Then, the goal was to control a flow of dynamical systems with attractors, see [Jackson, 1991]. For identical or complete synchronization of two systems, Pecora and Carroll presented a criterion of the sub-Lyapunov exponents to determine the synchronization of two systems connected with common signals. The common signals are as constraints for such two systems, [Pecora & Carroll, 1990-1991]. Since then, one focused on developing the corresponding control methods and schemes to achieve the synchronization of two dynamical systems with constraints.

In [Kapitaniak, 1994; Pyragas, 1992] some methods for chaos control with a small time continuous perturbation are presented, which can achieve a synchronization of two chaotic dynamical systems. In [Rulkov *et al.*, 1995] the author discuss a generalized synchronization of chaos in directionally coupled chaotic systems, exploring a condition, which equates dynamical variables from one subsystem with a function of the variables of another subsystem. In [Boccaletti *et al.*, 1997] an adaptive synchronization of chaos for secure communication was presented, an algorithm that provides synchronization between a message sender and a message receiver and assures security in the communication against external interceptions. In [Yang & Chua, 1999] the authors gives a study for a special case of generalized synchronization, in which the synchronization manifold is linear. In [Zhan *et al.*, 2003] the author show that generalized synchronization can be achieved by a single scalar signal, and its synchronization threshold for different delay times shows the parameter resonance effect, i.e., the stable synchronization is obtained at a smaller coupling if the delay time of the driven system is chosen such that it is in resonance with the driving system. Using statistical physics studies, the topology and dynamics of evolving of networks was investigated in [Dorogovtsev & Mendes, 2003]. The authors discuss the main models and analytical tools, covering random graphs, small-world and scale-free networks, the emerging theory of evolving networks, and the interplay between topology and the networks robustness against failures and attacks. It was shown in [Teufel *et al.*, 2006] that two different coupled pendulums can be synchronized by a fluid flow and derived conditions for the existence and stability of the synchronized motions depending on the coupling strength and the frequency detuning. Using normal form theory and the prevailing direct averaging approach the occurring Hopf bifurcation with two distinct pairs of purely imaginary eigenvalues is studied in the nonresonant case and in the 1:1-resonance corresponding, respectively, to strong and weak coupling. In [Stojanovski & Kocarev, 1997] the authors used symbolic dynamics and information theory to investigate the chaotic synchronization for discrete maps. Rulkov in [Rulkov, 2001] show how synchronization among chaotically bursting cells can lead to the onset of regular bursting. In order to present the mechanism behind such regularization, the author model the individual dynamics of each cell with a simple two-dimensional map that produces chaotic bursting behavior similar to biological neurons. In [Afraimovich *et al.*, 2002]the generalized chaos synchronization of non-invertible maps was investigated. They studied the properties of a functional relation between a non-invertible chaotic drive and a response map in the regime of generalized synchronization of chaos. It was shown in [Barreto *et al.*, 2003] that for coupled systems without symmetries, systems can be coherent without having easily-detectable synchronization properties. They have given examples of invertible and non-invertible drivers for which the system is asymptotically stable, yet the synchronization set is non-smooth or multi-valued. Furthermore, they give a method of detection designed for multi-fractal synchronization sets. The hybrid projective synchronization along a proposed systematic and concrete full state hybrid projective synchronization scheme for a general class of chaotic maps based on the active control idea was given in [Hu *et al.*, 2008]. Kuramoto in [Kuramoto, 2003] used the concept of phase synchronization to investigate the waves and turbulence in chemical oscillations. The author describes a few asymptotic methods that can be used to analyze the dynamics of self-oscillating fields of the reactive-diffusion type and some related systems and surveys some applications of them. In [Zaks *et al.*, 1999] the author studied the imperfect phase synchronization through the alternative locking ratios. In [Feng & Shen, 2005] the author investigate phase synchronization and anti-phase chaotic synchronization, in the generalized sense, for the degenerate optical parametric oscillator. In [Pareek *et al.*, 2005; Xiang *et al.*, 2008] multiple one-dimensional chaotic maps to were used to investigate cryptography. Secure communications using chaotic synchronization are also studied in several papers, such as in [Fallahi *et al.*, 2008] and in [Kiani *et al.*, 2009], by using Kalman filters and multi-shift cipher algorithm and in [Wang & Yu, 2009], by using a block encryption algorithm based on dynamic sequences of multiple chaotic systems. In [Soto & Akhmediev, 2005] it was shown that dissipative solitons can have dynamics similar to that of a strange attractor in low dimensional systems and using a model of a passively mode-locked fiber laser, it is shown that soliton pulsations with periods equal to several round-trips of the cavity can be chaotic, even though they are synchronized with the round-trip time. In [Hung *et al.*, 2006] it is studied the chaos synchronization of two stochastically coupled random Boolean networks. The coupling mechanism considered is that the $n^{th}$ cell in a network is linked by an arbitrarily chosen cell in the other network with probability $p$, and it possesses no links with probability $1-p$. This mechanism is useful to investigate the coevolution of biological species via horizontal genetic exchange. Synchronization on the dynamical systems with time-delay has been the subject of a recent very active investigation and some results can be found in [Bowong *et al.*, 2006], [Ghosh *et al.*, 2007], [Wang *et al.*, 2008] and [Cruz & Romero, 2008]. Outer synchronization is the synchronization between two discrete-time networks and it is theoretically and numerically studied in [Li *et al.*, 2009].

## 7.3. Some Physical Examples About the Real Applications of Network Synchronizability

The theory of complex networks offer an appropriate framework for the study of a large-scale and representative class of complex systems, with examples ranging from cell biology and epidemiology to the Internet, [Barabási & Albert, 1999; Watts & Strogatz, 1998; Strogatz, 2003]. The discovery of universal structural properties in real-world networks and the theoretical understanding of evolutionary laws governing the emergence of these properties has been used to the research in these areas. Most of biological, social and technological complex systems are inherently dynamic. Therefore, along with the study of purely structural and evolutionary properties, there has been increasing interest in the interplay between the dynamics and the structure of complex networks.

Epidemics and immunization are some of the applications of network synchronization. In [Satorras & Vespignani, 2003] a review of the main results obtained in the modelling of epidemics spreading in scale-free networks was provided. In particular, it was shown the different epidemiological framework originated by the lack of any epidemic threshold and how this feature is rooted in the extreme heterogeneity of the scale-free networks connectivity pattern. In [Tang *et al.*, 2009], the accumulation phenomenon in public places, was investigated how the condensation of moving bosonic particles influences the epidemic spreading in scale-free metapopulation networks. The mean-field theory shows that condensation can significantly enhance the effect of epidemic spreading and reduce the threshold for epidemic to survive, in contrast to the case without condensation. In the stationary state, the number of infected particles increases with the degree $k$ linearly when $k < k_c$ and nonlinearly when $k > k_c$, where $k_c$ denotes the crossover degree of the nodes with unity particle.

Neural networks are another application of network synchronization [Kinzel, 2003]. Modular organization is a special feature shared by many biological and social networks alike. It is a hallmark for systems exhibiting multitasking, in which individual tasks are performed by separated and yet coordinated functional groups. Understanding how networks of segregated modules develop to support coordinated multitasking functionalities is the main topic studied in [Fuchs *et al.*, 2009]. Using simulations of biologically inspired neuronal networks during development, is investigated the formation of functional groups (cliques) and inter-neuronal synchronization. The results indicate that synchronization cliques first develop locally according to the explicit network topological organization. Later on, at intermediate connectivity levels, when networks have both local segregation and long-range integration, new synchronization cliques with distinctive properties are formed. In particular, by defining a new measure of synchronization centrality, is identified at these developmental stages dominant neurons whose functional centrality largely exceeds the topological one. These are generated mainly in a few dominant clusters that become the centers of the newly formed synchronization cliques. It is shown that by the local synchronization properties at the very early developmental stages, it is possible to predict with high accuracy which clusters will become dominant in later stages of network development.

Transportation systems are complex dynamical systems whose dynamics may be described by networks. Traffic unfolds its dynamics on a graph and the dynamics on the links (roads) and nodes (intersections like train or bus stations), of this graph. Although this is also true for others networked systems, such as for electrical networks or for biological networks like nerve system and blood transport system, in the traffic dynamics on links there is a difference, since the agents or individuals have strategic long-term goals, [Nagel, 2003]. Economies of scale applied to transport modes and the expansion of transport infrastructure have increased the need for the synchronization of movements between transport terminals, and thus the synchronization of transport terminals themselves. This is the latest stage in a long process of transport system development.

The topology of a network often affects its functional behaviors. For instance, the topology of a social network affects the spreading of information and also disease, while an unsuitable topology of a power grid can damage its robustness and stability. In fact, we are confronting all kinds of networks with complex structures everyday, where handy instances are the Internet and the World Wide Web. The economic-cycle synchronous phenomenon in the World Trade Web is a scale-free type of social economic networks, used to illustrate an application of the network synchronization mechanism. It is important to take in account the way in which individuals decisions and actions are influenced by the network of connections that link them to other agents, [Kirman, 2003].

Cryptography and secure communications are also an application of network synchronization, [Pareek *et al.*, 2005], [Boccaletti *et al.*, 1997]. In recent years chaotic secure communication and chaos synchronization have been received more increasing attention, [Stavrinides *et al.*, 2009], [Wang & Yu, 2009], [Xiang *et al.*, 2008]. In [Fallahi *et al.*, 2008] a chaotic communication method using extended Kalman filter is presented. Encoding chaotic communication is used to achieve a satisfactory, typical secure communication scheme. In the proposed system, a multi-shift cipher algorithm is also used to enhance the security and the key cipher is chosen as one of the chaos states. The key estimate is employed to recover the primary data. A fractional chaotic communication method using an extended fractional Kalman filter may also be used, [Kiani *et al.*, 2009].

## 7.4. Preliminaries

In this chapter, we introduce some notions and basic results on graphs and discrete dynamical systems. Mathematically, networks are described by graphs (directed and undirected) and the theory of dynamical networks is a combination of graph theory and nonlinear dynamics. From the point of view of dynamical systems, we have a global dynamical system emerging from the interactions between the local dynamics of the individual elements. The graph theory allows us to analyze the coupling structure between them.

A graph $G$ is a set $G = (V, E)$ where $V = V(G)$ is a nonempty set of $N$ vertices or nodes ($N$ is called the order of the graph) and $E = E(G)(\subseteq V(G) \times V(G))$ is the set of $m$ pairs of vertices that are called edges or links $e_{ij}$ that connect two vertices $v_i$ and $v_j$. For a graph with $N$ vertices, the maximum cardinality of $E(G)$ is $\frac{N(N-1)}{2}$.

The matrix $A = A(G) = [a_{ij}]$, is called the adjacency matrix. For a non weighted graph, it carries an entry 1 at the intersection of the $j^{th}$ row and the $j^{th}$ column if there is a edge from $v_i$ to $v_j$, where $v_i, v_j \in V(G)$. When there is no edge, the entry will be 0. If the graph is not directed, $a_{ij} = a_{ji}$ and the matrix $A(G)$ is symmetric.

The degree of a node $v_i$, represented by $k_i$, is the number of edges incident on it, i.e.,

$$k_i = \sum_{\substack{j=1 \\ j \neq i}}^{N} a_{ij} \tag{7.1}$$

Consider the diagonal matrix $D = D(G) = [d_{ij}]$, where $d_{ii} = k_i$. We call Laplacian matrix to $L = D - A$. The eigenvalues of $L$ are all real and non negatives and are contained in the interval $[0, \min\{N, 2\Delta\}]$, where $\Delta$

is the maximum degree of the vertices. The spectrum of $L$ may be ordered, $\lambda_1 = 0 \leq \lambda_2 \leq \cdots \leq \lambda_N$. The second eigenvalue $\lambda_2$ is known as the algebraic connectivity or Fiedler value and plays a special role in the graph theory. As much larger $\lambda_2$ is, more difficult is to separate the graph in disconnected parts. The graph is connected if and only if $\lambda_2 \neq 0$. In fact, the multiplicity of the null eigenvalue $\lambda_1$ is equal to the number of connected components of the graph. As we will see later, as bigger is $\lambda_2$, more easily the network synchronizes.

Consider a network of $N$ identical chaotic dynamical oscillators, described by a connected graph, with no loops and no multiple edges. In each node the dynamics of the oscillators is defined by $\dot{x}_i = f(x_i)$, with $f : \mathbb{R}^n \to \mathbb{R}^n$ and $x_i \in \mathbb{R}^n$ is the state variables of the node $i$. The state equations of this network are

$$\dot{x}_i = f(x_i) + c \sum_{\substack{j=1 \\ j \neq i}}^{N} a_{ij} \Gamma(x_j - x_i), \;(i = 1, 2, ..., N) \tag{7.2}$$

where $c > 0$ is the coupling parameter, $A = [a_{ij}]$ is the adjacency matrix and $\Gamma = diag(1, 1, ...1)$. Eq. (7.2) can be rewritten as

$$\dot{x}_i = f(x_i) + c \sum_{j=1}^{N} l_{ij} x_j, \quad (i = 1, 2, ..., N) \tag{7.3}$$

where $L = (l_{ij}) = D - A$ is the Laplacian matrix or coupling configuration of the network. The network (7.3) achieves asymptotical synchronization if $x_1(t) = x_2(t) = ... = x_N(t) \underset{t \to \infty}{\to} e(t)$, where $e(t)$ is a solution of an isolate node (equilibrium point, periodic orbit or chaotic attractor), satisfying $\dot{e}(t) = f(e(t))$. Consider the network (7.3) with identical chaotic nodes in the discretized form

$$x_i(k+1) = f(x_i(k)) + c \sum_{j=1}^{N} l_{ij} f(x_j(k)) \tag{7.4}$$

with $i = 1, 2, ..., N$. It is known that a network (7.4) with identical chaotic nodes is synchronized if the coupling parameter $c$ belongs to the synchronization interval

$$\frac{1 - e^{-h_{\max}}}{|\lambda_2|} < c < \frac{1 + e^{-h_{\max}}}{|\lambda_N|} \tag{7.5}$$

where $\lambda_i$ $(i = 1, 2, ..., N)$ are the eigenvalues of the Laplacian $L$ and $h_{\max}$ is the Lyapunov exponent of each individual $n$-dimensional node.

Fixing the dynamics $f$ in the nodes, the synchronization interval will be as larger as much the eigenratio $r$

$$r = \frac{\lambda_2}{\lambda_N} \tag{7.6}$$

is bigger.

## 7.5. Complete Clustered Networks

Consider the graph $G = (V(G), E(G))$ associated to a network, as described above. In this section, among all graph invariants we will pay special attention to the clustering coefficient, the performance of a clustering and the conductance. A clustering of the graph $G$ is a partition of the vertices set $C^k = \{C_1, C_2, ..., C_k\}$ and the $C_i \subset V(G)$ are called clusters. $C$ is called trivial if either $k = 1$ or all clusters $C_i$ contain only one element. We identify a cluster $C_i$ with the induced subgraph of $G$, i.e., the graph $G(C_i) = \{C_i, E(C_i)\}$. If all clusters in a clustering are complete graphs (cliques), we call it a complete clustered network. For each fixed clustering $C^k = \{C_1, C_2, ..., C_k\}$ the set of intracluster edges and the set of intercluster edges are defined, respectively, by

$$Intra(C) = \bigcup_{i=1}^{k} E(C_i) \quad \text{and} \tag{7.7}$$

$$Inter(C) = E(G) \backslash Intra(C) \tag{7.8}$$

Given a clustering, remains now the question about its quality. The partition of the set of vertices can be arbitrary, however, not all partitions are what we think it is an intuitively clustering. A good clustering should be a partition with a large number of intracluster edges and few intercluster edges. Of course, if we have a complete graph, any non-trivial clustering will be a bad clustering. Conversely, if the graph is disconnected with $k$ components, the partition $C^k$ corresponding to these components is a good clustering. But what is the point separating the two possibilities? Trying to answer this question we introduce some known graph parameters. We begin with the clustering coefficient.

### 7.5.1. *Clustering Point on Complete Clustered Networks*

The clustering coefficient measures the number of mutual neighbors of adjacent nodes, such that, the average probability for two neighbors of some vertex to be directly connected. The clustering coefficient $c_i$ of a vertex $v_i$ is given by the proportion of edges between the vertices within its neighborhood divided by the number of edges that could possibly exist between them. The clustering coefficient $c_i$ of a vertex $v_i$, with degree larger than one, of a graph, is defined, [Watts & Strogatz, 1998], by

$$c_i = \frac{|\{e_{jk}\}|}{\binom{k_i}{2}} = \frac{2\,|\{e_{jk}\}|}{k_i(k_i-1)}, \text{ with } e_{jk} \in E;\ v_j, v_k \in N_i \tag{7.9}$$

where $N_i = \{v_j : e_{ij} \in E\}$ is the set of the neighbors of vertex $v_i$.

In terms of the adjacency matrix $A = A(G) = [a_{ij}]$ the clustering coefficient can be calculated by

$$c_i = \frac{\sum\limits_{j=1}^{N} \sum\limits_{m=1}^{N} a_{ij} a_{jm} a_{mi}}{k_i(k_i-1)} \tag{7.10}$$

This was the local aspect of clustering coefficient. To consider the global aspect, which is also called community structure, it is necessary to look for groups of vertices with many connections within a group, but considerably fewer between groups. That parameter characterizes how densely clustered the edges in a network are. To characterize the global clustering coefficient of the network we introduce the average of $c_i$ over all vertices with degree larger than one

$$c = \frac{\sum\limits_{i,k_i>1} c_i}{\sum\limits_{i,k_i>1} 1} \tag{7.11}$$

To test the behavior of this parameter in the formation of clusters, we simulate the creation of clusters starting with a complete graph of $N = 15$ vertices where each vertex is connected to every other one, excluding self-connections. We present three different situations, a $C^3$, a $C^4$ and a $C^5$ clustering. In each one we begin with a complete graph, that means with $\frac{N(N-1)}{2} = 105$ edges, and by deleting intercluster edges leads to the creation of three, four and five clusters. From the complete graph we delete edges until the disconnection of the graph by the formation of the complete clustered network. Before that point is already evident the appearance of the clustering. What should be the moment we consider the graph clustered?

What is the number of edges where the formation of clusters is clear? Let us observe the three examples.

For the case where we obtain three clusters, $C^3 = \{\{1,2,3,4,5\}, \{6,7,8,9\}, \{10,11,12,13,14,15\}\}$, Fig. 7.1 shows four steps of the evolution of the network. The Fig. 7.2 shows the behavior of the clustering

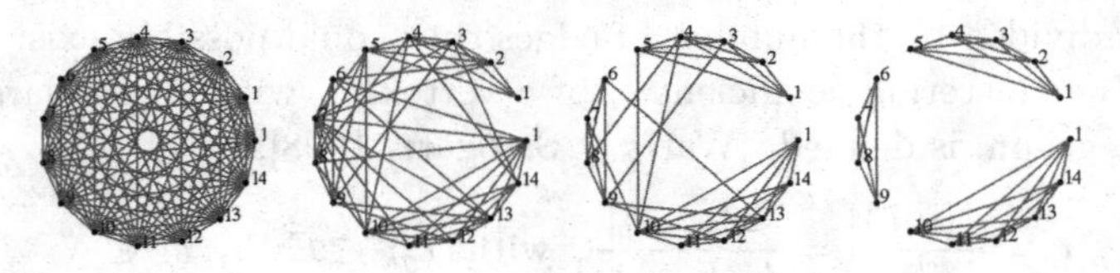

Fig. 7.1. Formation process of three clusters.

coefficient $c$, on the formation of the three clusters $C^3$. Note that, in this

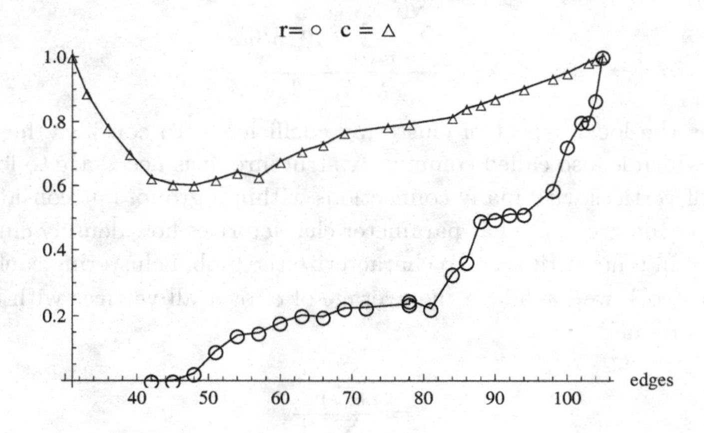


Fig. 7.2. Evolution of the clustering coefficient $c$ and the eigenratio $r$ in the formation process of the $C^3$ clustering.

example, $N = |V(G)| = 15$, $|Intra(C^3)| = 31$ and we observe that the formation of clusters is apparent when $|E(G)| < 47$. The Fig. 7.3 shows four steps of the evolution of the network to a $C^4$ clustering. We begin with a complete graph (with 105 edges) and after deleting edges we obtain four clusters, $C^4 = \{\{1,2,3\}, \{4,5,6,7\}, \{8,9,10\}, \{11,12,13,14,15\}\}$. The Fig. 7.4 shows the behavior of the clustering coefficient $c$, on the formation of the four clusters $C^4$. For this case, $N = 15$ and $|Intra(C^4)| = 22$ and we observe that the formation of clusters is apparent when $|E(G)| < 38$. For

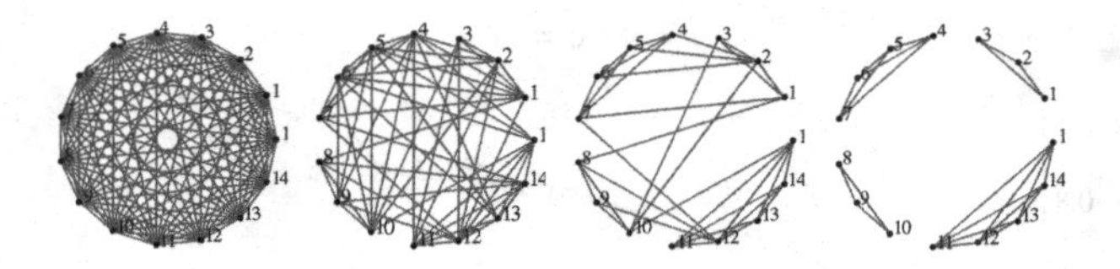

Fig. 7.3. Formation process of four clusters.

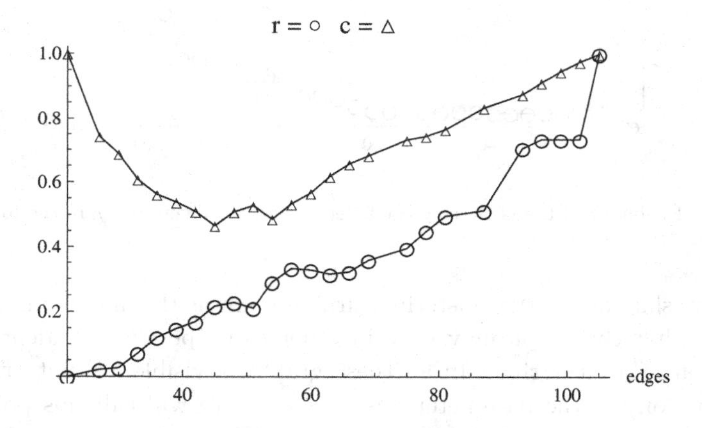


Fig. 7.4. Evolution of the clustering coefficient $c$ and the eigenratio $r$ in the formation process of the $C^4$ clustering.

the case where we obtain five clusters $C^5 = \{\{1, 2, 3\}, \{4, 5, 6\}, \{7, 8, 9\}, \{10, 11, 12\}, \{13, 14, 15\}\}$ the Fig. 7.5 shows four steps of the evolution of the network: we start, again, with a complete graph (with 105 edges) and after deleting edges we obtain five clusters. The Fig. 7.6 shows the behav-

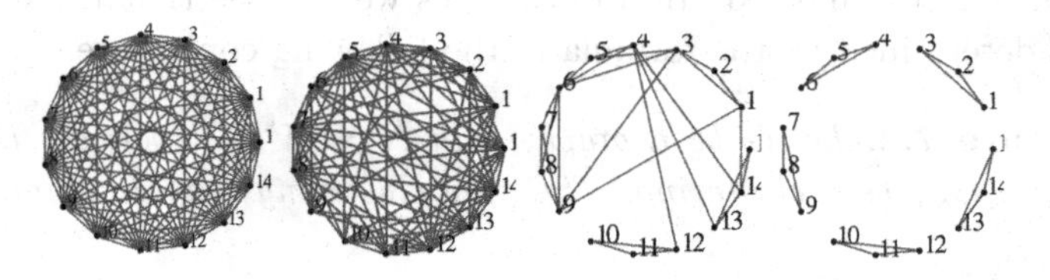

Fig. 7.5. Formation process of five clusters.

ior of the clustering coefficient $c$, on the formation of the five clusters $C^5$. In such a clustering process, starting with a complete graph and deleting

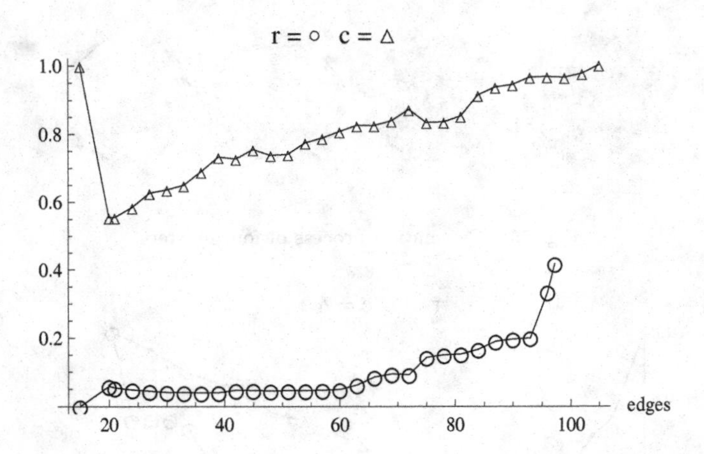


Fig. 7.6. Evolution of the clustering coefficient $c$ and the eigenratio $r$ in the formation process of the $C^5$ clustering.

edges, we shall define the clustering turning point as the maximum number of edges that the graph may have in order to be possible to identify the formation of a clustering. In all these graphs is visible a point at which the monotony of the parameter $c$ is changed. We will call this point the clustering turning point. Obviously, it should depend on the number of vertices but also in the way the edges are removed.

**Conjecture 7.1.** *In a clustering process, starting with a complete graph and deleting edges, the clustering coefficient decreases before a certain point and inverts its monotonicity after that point. In this process, the synchronizability decreases always.*

See Figs. 7.2, 7.4 and 7.6. In all examples we can see that the clustering point is determined by the formula in the following conjecture.

**Conjecture 7.2.** *Let $G$ be a graph, with $N = |V(G)|$ vertices let $C^k = \{C_1, C_2, ..., C_k\}$ be a clustering. Then the clustering turning point is given by:*

$$Cl(C^k) = N + \left|Intra(C^k)\right| + 1 \tag{7.12}$$

You can say that the claim "*a large value of the clustering coefficient enhances the synchronization*", [Comellas & Gago, 2007], is true for example, in the case one delete edges following a certain path and in the beginning of cluster formation, but is not true after the clusters appearance, [Caneco *et al.*, 2009].

### 7.5.2. *Classification of the Clustering and the Amplitude of the Synchronization Interval*

In the context of random walks in graphs, the conductance appeared as a measure of fluidity, being smaller with the existence of certain parts of the state space which retained the process (funnels). Techniques to know how long must the chain evolve to attempt some given proximity of the stationary distribution are of great importance in problems like coupling and synchronization. Intuitively, a random walk is rapidly mixing, that is, within a polynomial number of steps, in the problem size, the Markov chain approaches the equilibrium, if there are no bottlenecks or funnels.

There are several definitions of conductance, [Fernandes *et al.*, 2010], we will use the following

$$\phi(G) = \min_{U \subset V} \frac{|E(U, V-U)|}{\min\{|E_1(U)|, |E_1(V-U)|\}} \tag{7.13}$$

In this definition, $|E(U, V-U)|$ is the number of edges from $U$ to $V-U$ and $|E_1(U)|$ means the sum of degrees of vertices in $U$.

The performance of a clustering should measure the quality of each cluster as well as the cost of the clustering. In [Kannan *et al.*, 2004] this bicriteria is based in a two-parameter definition of a $(\alpha, \varepsilon)$-clustering, where $\alpha$ should measure the quality of the clusters and $\varepsilon$ the cost of such partition, that is, the ratio of the intercluster edges to the total of edges in the graph.

**Definition 7.1.** A partition $C^k = \{C_1, C_2, ..., C_k\}$ of $V$ is an $(\alpha, \varepsilon)$-clustering if:

(1) The conductance of each cluster is at least $\alpha$

$$\Phi(G(C_i)) \geq \alpha, \text{ for all } i = 1, ..., k;$$

(2) The fraction of intercluster edges to the total of edges is at most $\varepsilon$

$$\frac{Inter\,(C^k)}{|E(C^k)|} \leq \varepsilon$$

According to this definition, the clustering is good if it maximizes $\alpha$ and minimizes $\varepsilon$. We introduce then a coefficient that accomplish both optimization problems.

**Definition 7.2.** For an $(\alpha, \varepsilon)$-clustering $C$, define the performance of $C$ by the ratio

$$R = \frac{\varepsilon}{\alpha}$$

That means that a clustering is better if it has smaller $R$. We simulate again the creation of clusters starting with a complete graph of $N$ vertices and deleting edges leading to the creation of three, four and five clusters. For the case where we obtain three clusters $C^3 = \{\{1,2,3,4,5\}, \{6,7,8,9\}, \{10,11,12,13,14,15\}\}$ the Fig. 7.1 shows four steps of the evolution of the network: we start with a 15−node complete graph, with 105 edges.

The Figs. 7.7 and 7.8 show the behavior of the graph invariants: the conductance $\phi$, the performance $R$ and the eigenratio $r$; the clustering coefficient $c$, the conductance $\phi$ and the eigenratio $r$, respectively, on the formation of the three clusters $C^3$.

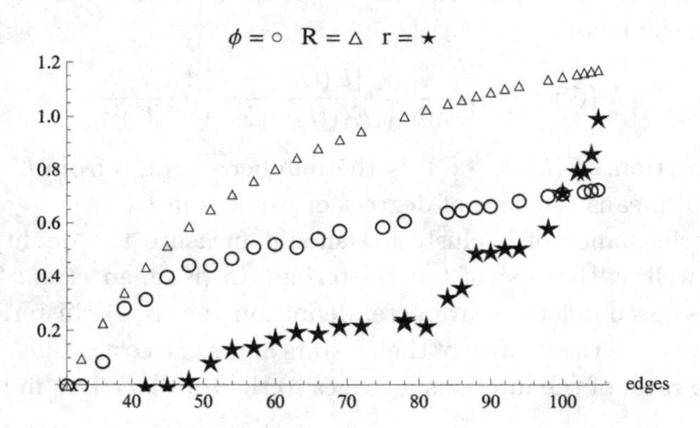


Fig. 7.7. Evolution of the conductance $\phi$, the performance $R$ and the eigenratio $r$ in the formation process of the $C^3$ clustering.

Fig. 7.3 shows four steps of the evolution of the network: we begin with a complete graph (with 105 edges) and after deleting edges we obtain four clusters, $C^4 = \{\{1,2,3\}, \{4,5,6,7\}, \{8,9,10\}, \{11,12,13,14,15\}\}$.

The Fig. 7.9 shows the behavior of the graph invariants: the conductance $\phi$, the performance $R$ and the eigenratio $r$, on the formation of the four clusters $C^4$. Fig. 7.10 shows the results that compare the parameters clustering coefficient $c$, conductance $\phi$ and eigenratio $r$.

For the case where we obtain five clusters $C^5 = \{\{1,2,3\}, \{4,5,6\}, \{7,8,9\}, \{10,11,12\}, \{13,14,15\}\}$ in Fig. 7.5 is shown four steps of the

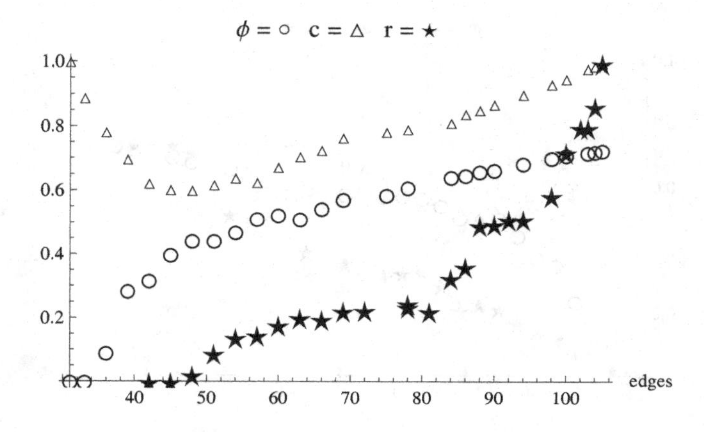


Fig. 7.8. Evolution of the clustering coefficient $c$, the conductance $\phi$, and the eigenratio $r$ in the formation process of the $C^3$ clustering.

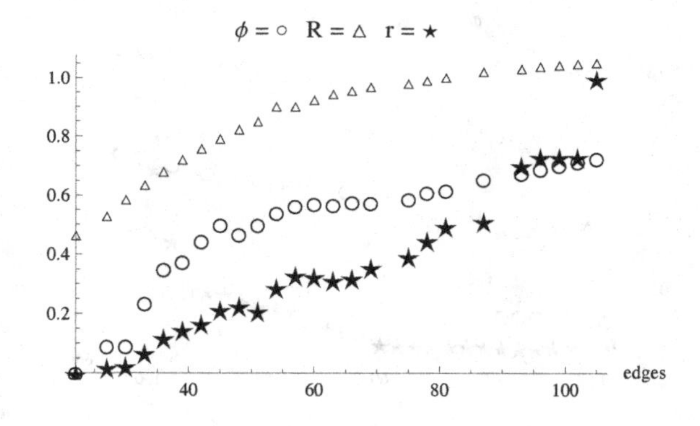


Fig. 7.9. Evolution of the conductance $\phi$, the performance $R$ and the eigenratio $r$ in the formation process of the $C^4$ clustering.

evolution of the network: we start, again, with a complete graph (with 105 edges) and after deleting edges we obtain five clusters. The Figs. 7.11 and 7.12 show the behavior of the graph invariants: the conductance $\phi$, the performance $R$ and the eigenratio $r$; the clustering coefficient $c$, the conductance $\phi$ and the eigenratio $r$, respectively, on the formation of the five clusters $C^5$.

We can observe, that the conductance enables us to evaluate the synchronizability of the network. Low conductance means that there is some bottleneck in the graph, a subset of nodes not well connected (a few in-

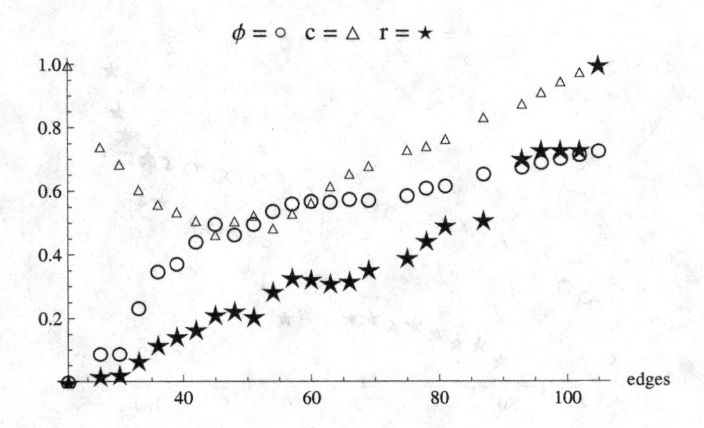


Fig. 7.10. Evolution of the clustering coefficient $c$, the conductance $\phi$, and the eigenratio $r$ in the formation process of the $C^4$ clustering.

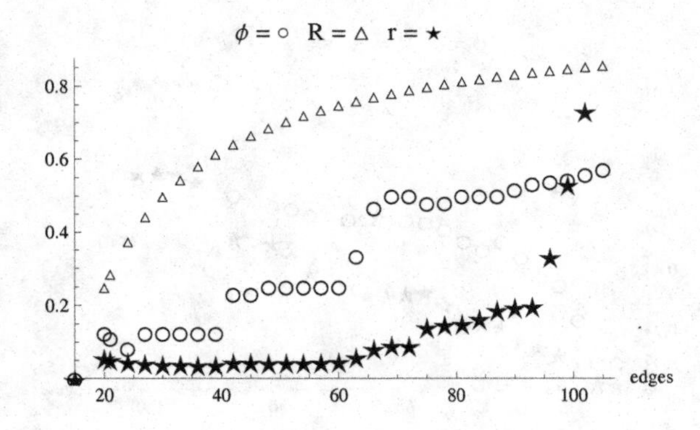


Fig. 7.11. Evolution of the conductance $\phi$, the performance $R$ and the eigenratio $r$ in the formation process of the $C^5$ clustering.

terclusters edges) with the rest of the graph. With these simulations it is possible to establish the following result.

**Conjecture 7.3.** *A bad performance of the clustering implies a larger synchronization interval and a high conductance.*

### 7.5.3. *Discussion*

In this section, we conjectured some results about the formation and identification of clusterings in a network. All numerical evidences were done

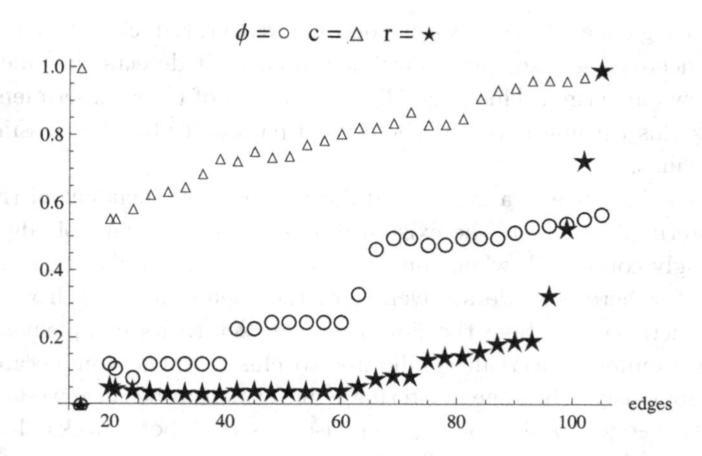


Fig. 7.12. Evolution of the clustering coefficient $c$, the conductance $\phi$, and the eigenratio $r$ in the formation process of the $C^5$ clustering.

for a graph with 15 nodes, giving rise to what we call complete clustered networks, in the three following different cases $C^3$, $C^4$ and $C^5$:

$$\{\{1,2,3,4,5\},\{6,7,8,9\},\{10,11,12,13,14,15\}\}$$

$$\{\{1,2,3\},\{4,5,6,7\},\{8,9,10\},\{11,12,13,14,15\}\}$$

$$\{\{1,2,3\},\{4,5,6\},\{7,8,9\},\{10,11,12\},\{13,14,15\}\}$$

In all three cases we computed the values of the clustering coefficient, the conductance, the amplitude of the synchronization interval and the performance of the considered clusters. There are big differences between the meaning of each such parameter and we believe that all of them have a particular interest.

The clustering coefficient is independent of a given clustering, it is a property of the graph and evaluates the density of edges around each vertex. This lets you know if it is convenient to divide the graph into clusters or not. Our experiments shows that there exists a moment in the process of formation of the clustering, where the monotonicity of the clustering coefficient changes. When we delete edges it tends to decrease but it is true until a certain point after which it increases. Of course, if we have a complete graph, is natural that the clustering coefficient $c$ is high, since the links between the vertices are the strongest possible. But this high degree of clustering refers to the entire graph as the trivial clustering. When we begin erasing edges the connections become weaker, reducing the coefficient $c$. This decreasing takes place until the moment it detects the emergence of the new clustering. This is highlighted in each of three experiments and we think this turning point will be a good parameter for the classification of clusterings.

The conductance of a graph is also a property independent of the chosen clustering. It detects the existence of a group or groups of edges that are strongly connected, when compared with others. Unlike the clustering coefficient, where is made an average of the coefficient of each vertex, in the conductance, we take the flow from one set to its complement, and take the minimum over all candidates to cluster. The conductance decreases obviously, when the clustering is in formation because we force the intercluster edges to disappear giving rise to strong bottlenecks. It is also understandable that a slower fluidity in the system gives rise to a slower synchronization. We also compared it with the amplitude of the synchronization interval and the result was as expected. However we have no formal relation between synchronization and conductance.

The only parameter that evaluates the performance of a given clustering is the parameter $R = \varepsilon/\alpha$. This ratio depends on the considered clustering and varies with it. Again, there is not a formalism that allows us to establish a formal relationship between the synchronization and quality of a clustering, despite the numerical evidence.

## 7.6. Symbolic Dynamics and Networks Synchronization

The use of symbolic dynamics to study network synchronization is a recent approach. From the point of view of dynamical systems we have a global dynamical system emerging from the interactions between the local dynamics of the individual elements. Symbolic dynamics is a powerful tool to study and understand not only the behavior of the local dynamical systems in the nodes, but also the global dynamics in the network. The main idea is to encode the itineraries of the images of the critical (or discontinuity) points of a piecewise monotone map $f$ on an interval in symbolic sequences. These symbolic sequences are called kneading sequences, after [Milnor & Thurston, 1988]. The set of kneading sequences for a map $f$ is called the kneading invariant of $f$ and is denoted by $\mathcal{K}(f)$. This kneading invariant is a topological invariant and can be used to obtain topological information on the discrete dynamical system induced by $f$, such as the topological entropy, $h_{top}(f)$. See [Fernandes & Ramos, 2006], [Lampreia & Ramos,

1997], [Milnor & Thurston, 1988], [Rocha & Ramos, 2006] and references therein. Consider a compact interval $I \subset \mathbb{R}$ and a $m$-modal map $f : I \to I$, i. e., the map $f$ is piecewise monotone, with $m$ critical points and $m+1$ subintervals of monotonicity. Suppose $I = [c_0, c_{m+1}]$ can be divided by a partition of points $\mathcal{P} = \{c_0, c_1, ..., c_{m+1}\}$ in a finite number of subintervals

$$I_1 = [c_0, c_1], \quad I_2 = [c_1, c_2], \quad ... \quad I_{m+1} = [c_m, c_{m+1}]$$

in such a way that the restriction of $f$ to each interval $I_j$ is strictly monotone, either increasing or decreasing. Assuming that each interval $I_j$ is the maximal interval where the function is strictly monotone, these intervals $I_j$ are called laps of $f$ and the number of distinct laps is called the lap number, $\ell$, of $f$. In the interior of the interval $I$ the points $c_1$, $c_2$, ..., $c_m$, are local minimum or local maximum of $f$ and are called turning or critical points of the function. The limit of the $n$-root of the lap number of $f^n$ (where $f^n$ denotes the composition of $f$ with itself $n$ times) is called the growth number of $f$, and its logarithm is the topological entropy, denoted by

$$s = \lim_{n \to \infty} \sqrt[n]{\ell(f^n)} \quad \text{and} \quad h_{top} = \log s$$

The intervals $I_j = [c_{j-1}, c_j]$ are separated by the critical points, numbered by its natural order $c_1 < c_2 < ... < c_m$. We compute the images by $f$, $f^2$, ..., $f^n, ...$ of a critical point $c_j$ ($j = 1, ..., m-1$) and we obtain its orbit

$$O(c_j) = \{c_j^n : c_j^n = f^n(c_j),\ n \in \mathbb{N}\}$$

If $f^n(c_j)$ belongs to an open interval $I_k = ]c_{k-1}, c_k[$, then we associate to it a symbol $L_k$, with $k = 1, ..., m+1$. If there is an $r$ such that $f^n(c_j) = c_r$, with $r = 1, ..., m$, then we associate to it the symbol $A_r$. So, to each point $c_j$, we associate a symbolic sequence, called the address of $f^n(c_j)$, denoted by $S = S_1 S_2 ... S_n ...$, where the symbols $S_k$ belong to the $m$-modal alphabet, with $2m+1$ symbols, *i.e.*,

$$\mathcal{A}_m = \{L_1, A_1, L_2, A_2, ..., A_m, L_{m+1}\}.$$

The symbolic sequence $S = S_0 S_1 S_2 ... S_n ...$, can be periodic, eventually periodic or aperiodic. The address of a critical point $c_j$ is said eventually periodic if there is a number $p \in \mathbb{N}$, such that the address of $f^n(c_j)$ is equal to the address of $f^{n+p}(c_j)$, for large $n \in \mathbb{N}$. The smallest of such $p$ is called the eventual period. To each symbol $L_k \in \mathcal{A}_m$, with $k = 1, ..., m+1$ define its sign by

$$\varepsilon(L_k) = \begin{cases} -1 \text{ if } f \text{ is decreasing in } I_k \\ 1 \text{ if } f \text{ is increasing in } I_k \end{cases} \tag{7.14}$$

and $\varepsilon(A_k) = 0$, with $k = 1, ..., m$. We can compute the numbers

$$\tau_k = \prod_{i=0}^{k-1} \varepsilon(L_k)$$

for $k > 0$, and take $\tau_0 = 1$. The invariant coordinate of the symbolic sequence $S$, associated with a critical point $c_j$, is defined as the formal power series

$$\theta_{c_j}(t) = \sum_{k=0}^{k=\infty} \tau_k \, t^k \, S_k$$

The kneading increments of each critical point $c_j$ are defined by

$$\nu_{c_j}(t) = \theta_{c_j^+}(t) - \theta_{c_j^-}(t) \quad \text{with} \quad j = 1, ...m$$

where $\theta_{c_j^{\pm}}(t) = \lim_{x \to c_j^{\pm}} \theta_x(t)$.

Separating the terms associated with the symbols $L_1$, $L_2$, ...,$L_{m+1}$ of the alphabet $\mathcal{A}_m$, the increments $\nu_j(t)$, are written in the form

$$\nu_{c_j}(t) = N_{j1}(t)L_1 + N_{j2}(t)L_2 + \; ... \; + N_{j(m+1)}(t)L_{m+1}$$

The coefficients $N_{jk}$ in the ring $Z\,[[t]]$ are the entries of the $m \times (m+1)$ kneading matrix

$$N(t) = \begin{bmatrix} N_{11}(t) & \cdots & N_{1(m+1)}(t) \\ \vdots & \ddots & \vdots \\ N_{m1}(t) & \cdots & N_{m(m+1)}(t) \end{bmatrix}$$

From this matrix we compute the determinants $D_j(t) = det\,\widehat{N}(t)$, where $\widehat{N}(t)$ is obtained from $N(t)$ removing the $j$ column ($j = 1, ..., m+1$), and

$$D(t) = \frac{(-1)^{j+1} D_j(t)}{1 - \varepsilon(L_j)t}$$

is called the kneading determinant. Here $\varepsilon(L_j)$ is defined like in (7.14).

Let $f$ be a $m$-modal map and $D(t)$ defined as above. Let $s$ be the growth number of $f$, then the topological entropy of the map $f$ is, see [Milnor & Thurston, 1988],

$$h_{top}(f) = \log s, \quad \text{with} \quad s = \frac{1}{t^*} \quad \text{and} \tag{7.15}$$

$$t^* = \min\{t \in [0,1] : \; D(t) = 0\} \tag{7.16}$$

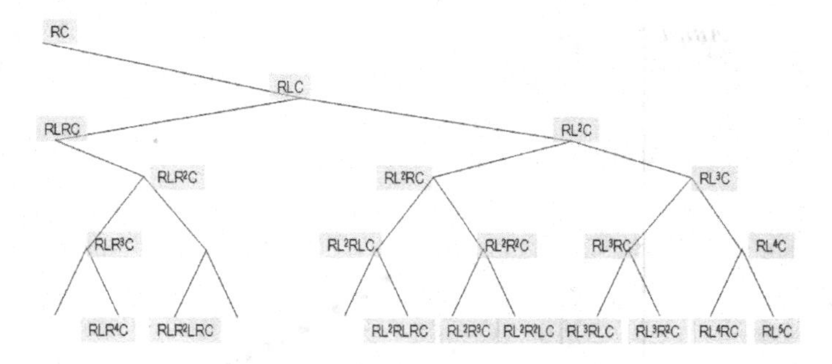


Fig. 7.13. Tree of unimodal kneading sequences.

We will use the symbolic dynamics to obtain results about a large set of functions $f$, namely those associated with the unimodal kneading sequences in the ordered tree in Fig. 7.13, [Lampreia & Ramos, 1997].

Sousa Ramos and some of his students have worked extensively this tree and obtained several results concerning the invariants that characterize the different dynamics present in each node of this tree. Using techniques from symbolic dynamics one can compute topological entropy, conductance, mixing rate, Lyapunov exponents, Hausdorff dimension, escape rate and others topological and metrical invariants. Our first approach consists in determining for each one of the local dynamic system for what values of the coupling parameter, $c$, there is synchronization, for different coupling scenarios.

Consider the network described by Eq. (7.4), where the function $f$, representing the local dynamics, is a piecewise linear map with slope $\pm s$ everywhere, i.e., topological entropy $h_{top} = \log s$, [Milnor & Thurston, 1988]. We denote such functions by $f_s$.

In this context, using the result from [Li & Chen, 2003] and attending that $h_{\max} = \log s$, [Rocha & Ramos, 2006], we are in position to state that the synchronization interval of these networks may be expressed in terms of the topological entropy. Fix some topology for the network, described by the adjacency matrix $A$ and vary the local map $f$ in the nodes, along the most right line, in the descent way, of the tree in Fig. 7.13, of admissible trajectories for the unimodal map. In this line the entropy grows and in Fig. 7.14 we see that the amplitude of the synchronization interval decreases as the entropy grows along this line.

The same result is obtained if we follow any horizontal line of this tree, going from the left to right. The amplitude of the interval decreases as the topological entropy grows.

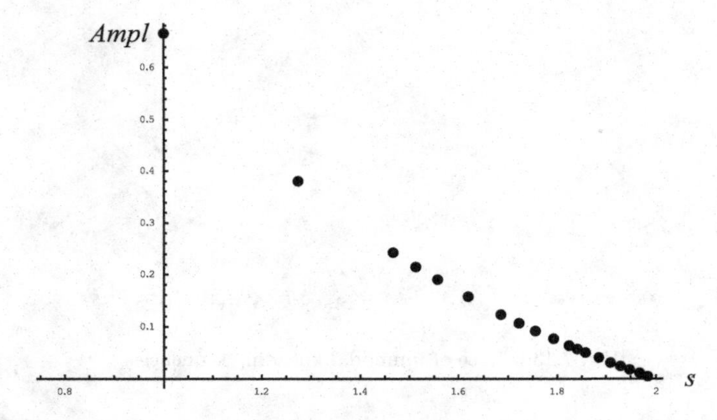


Fig. 7.14. The amplitude of the synchronization interval decreases when the topological entropy grows.

In Figs. 7.15, 7.16 and 7.17 we can observe the behavior of the synchronization interval for networks with fixed topology, given by the graph (or the matrix A) and in each node maps $f_s$, ordered according to the growth rate $s$. In these figures, as $s$ grows, the descendent line represents the upper limit of the synchronization interval and the ascendent line represents the lower limit. The amplitude of the synchronization interval, for each value of $s$, is the vertical distance between the two lines. So, if for a certain $s_0$, the two lines cross each other, then for $s > s_0$ the synchronization interval is empty. Note that in Fig.7.16 the graph is fully connected, then the synchronization interval of $f_s$ is nonempty for all values of $s$. In all figures we see that the amplitude of the synchronization interval decreases with $s$.

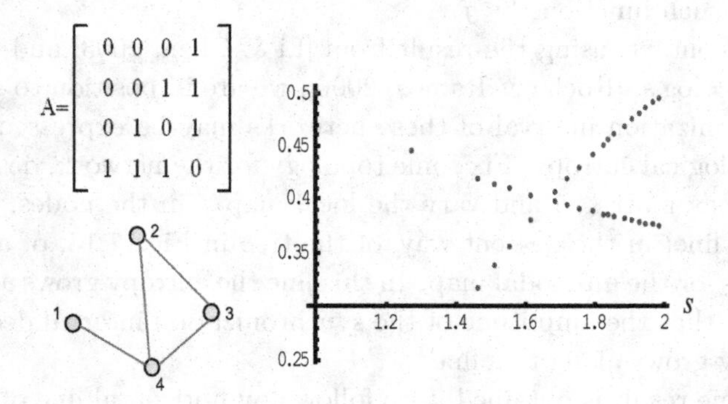


Fig. 7.15. Adjacency matrix and graph of the network and synchronization interval of $f_s$.

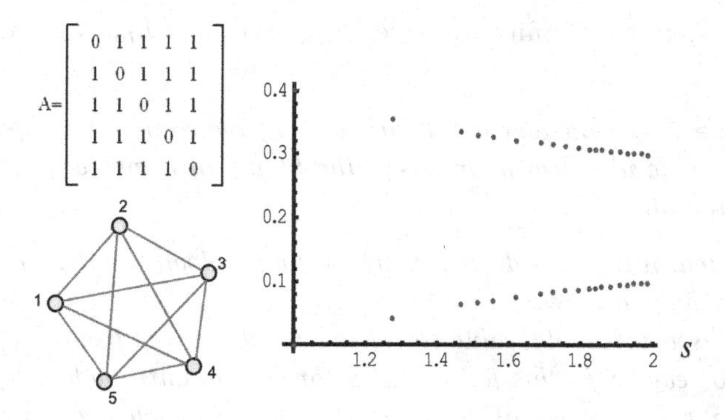


Fig. 7.16. Adjacency matrix, fully connected graph and synchronization interval of $f_s$.

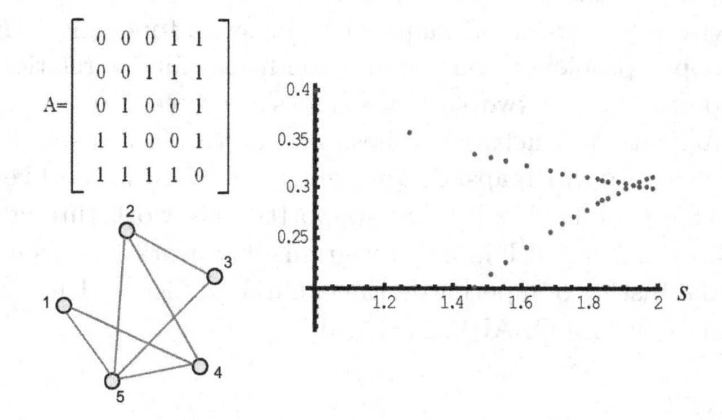


Fig. 7.17. Adjacency matrix and graph of the network and synchronization interval of $f_s$.

We studied networks having in each node identical chaotic piecewise linear maps. We obtained a synchronization interval in terms of the topological entropy in each node and in terms of the eigenvalues of the coupling matrix A, which determines the topology of the networks.

**Proposition 7.1.** *The synchronizability in the family of piecewise linear unimodal maps decreases when the topological entropy $h_{top}(f)$ in each node increases.*

***Proof.*** From (7.5) we have the synchronization interval given by

$$I = \left]\frac{1 - e^{-h_{\max}}}{|\lambda_2|}, \frac{1 + e^{-h_{\max}}}{|\lambda_N|}\right[$$

For a map $f_s$ in this family we have $h_{\max} = \log s = h_{top}(f)$, so the result follows. □

**Conjecture 7.4.** *Consider a network with a fixed connection topology, having in each node identical members of the family of m-modal functions f in the real interval. Then,*

*i) The synchronizability decreases when the topological entropy $h_{top}(f)$, in each node, increases.*

*ii) There is a threshold value $h_0$ of the topological entropy $h_{top}(f)$, such that, for each f in this family, with topological entropy $h < h_0$, the synchronization interval of the network with f in each node is nonempty.*

We have yet no proof for this conjecture but it should be possible to prove it, using the existing semiconjugacy between arbitrary m-modal maps and piecewise linear $m$-modal maps with the same topological entropy. It remains an open problem to find weaker conditions for the relation between the synchronization in networks whose nodes are piecewise monotone maps and synchronization in networks whose nodes are the respective semiconjugated piecewise linear maps. Acknowledgments: We would like to thank FCT (Portugal) for having in part supported this work through the Research Units Pluriannual Funding Program. The work has been also supported by the Instituto Superior de Engenharia de Lisboa, Universidade de Évora, CIMA-UE and CEAUL, Portugal.

## References

1. Afraimovich, V., Cordonet, A. & Rulkov, N. F. [2002] "Generalized synchronization of chaos in noninvertible maps," *Phys. Rev. E* **66**, 016208–1–6.
2. Barabási, A. L. & Albert, R. [1999] "Emergence of scaling in random networks," *Science* **286**, 509–512.
3. Barreto, E., Josic, K., Morales, C., Sander, E. & So, P. [2003] "The geometry of chaos synchronization," *Chaos* **13**, 151–164.
4. Boccaletti, S., Latora, V., Moreno, Y., Chavez, M. & Hwang, D. U. [2006] "Complex networks: Structure and dynamics," *Phys. Rep* **424**, 175–308.
5. Boccaletti, S. [2008] *The synchronized dynamics of complex systems*, Amsterdam, Elsevier.
6. Boccaletti, S., Farini, A. & Arecchi, F. T. [1997] "Adaptive synchronization of chaos for secure communication," *Phys. Rev. E* **55**, 4979–4981.
7. Bollobás, B. [1985] *Random Graphs*, New York.
8. Bowong, S., Moukam, K. F. M., Dimi, J. L. & Koina, R. [2006] "Synchronizing chaotic dynamics with uncertainties using a predictable synchronization delay design," *Commun. Nonlinear Sci. Numer. Simulat* **11**, 973–987.

9. Bezrukov, S. L. [2003] "Edge Isoperimetric Problems on Graphs," *Theor. Comp. Sci* **307**, 473–492.
10. Caneco, A., Grácio, C. & Rocha, J. L. [2008] "Symbolic dynamics and chaotic synchronization in coupled Duffing oscillators," J. *Nonlinear Math. Phys* **15**, 3, 102–111.
11. Caneco, A., Grácio, C. & Rocha, J. L. [2009] "Topological entropy for the sinchronization of piecewise linear and monotone maps. Coupled Duffing oscillators," *Int. J. Bifurcation & Chaos* **19**, 11, 3855–3868.
12. Caneco, A., Fernandes, S., Grácio C. & Rocha, J. L. [2010] "Networks Synchronizability, Local Dynamics and Some Graph Invariants," *Dynamics, Games and Science*, Peixoto, M. M., Pinto, A. A. and Rand, D. A. J. (eds), Springer (in press).
13. Comellas, F. & Gago, S. [2007] "Synchronizability of complex networks," *J. Phys. A: Math. Theor* **40**, 4483–4492.
14. Cruz-H. C., & Romero-H. N. [2008] "Communicating via synchronized time-delay Chua circuits," *Commun. Nonlinear Sci. Numer. Simulat* **13**, 645–659.
15. Dorogovtsev, S. N. & Mendes, J. F. F. [2003] *Evolution of Networks: From biological networks to the Internet and WWW*, Oxford University Press.
16. Fallahi, K., Raoufi, R. & Khoshbin, H. [2008] "An application of Chen system for secure chaotic communication based on extended Kalman filter and multi-shift cipher algorithm," *Commun. Nonlinear Sci. Numer. Simulat* **13**, 763–781.
17. Feng, J., Jost, J. & Qian, M. [2007] *Networks: From Biology to Theory*, Springer.
18. Feng, X. Q. & Shen, K. [2005] "Phase synchronization and anti-phase synchronization of chaos for degenerate optical parametric oscillator," *Chin. Phys* **14**, 1526–1532.
19. Fernandes, S. & Ramos, J. S. [2006] "Conductance, Laplacian and mixing rate in discrete dynamical systems," *Nonlinear. Dyn* **44**, 117–125.
20. Fernandes, S., Grácio, C. & Ramos, C. [2010] "Conductance in Discrete Dynamical Systems," *Nonlinear. Dyn* (in press).
21. Fuchs, E., Ayali, A., Ben-Jacob, E. & Boccaletti, S. [2009] "The formation of synchronization cliques during the development of modular neural networks," *Phys. Biol* **6**, 1–12.
22. Ghosh, D., Saha, P., Roy, C. A. [2007] "On synchronization of a forced delay dynamical system via the Galerkin approximation," *Commun. Nonlinear Sci. Numer. Simulat* **12**, 928–941.
23. Hu, M., Xu, Z. & Zhang, R. [2008] "Full state hybrid projective synchronization of a general class of chaotic maps," *Commun. Nonlinear Sci. Numer. Simulat* **13**, 782–789.
24. Hung, Y. -C., Ho, M. C., Lih, J. -S. & Jiang, I. -M. [2006] "Chaos synchronization of two stochastically coupled random Boolean networks," *Phys. Lett. A* **356**, 35–43.
25. Jackson, E. A. [1991] "Controls of dynamic flows with attractors," *Phys. Rev. E* **44**, 4839–4853.

26. Li, C., Xu, C., Sun, W., Xu, J. & Kurths, J. [2009] "Outer synchronization of coupled discrete-time networks," *Chaos* **19**, 013106–1–7.
27. Li, X. & Chen, G. [2003] "Synchronization and desynchronization of complex dynamical networks: An engineering viewpoint," *IEEE Trans. on Circ. Syst. -I* **50**, 1381–1390.
28. Kannan, R., Vempala, S. & Vetta, A. [2004] "On Clusterings: Good, bad and spectral," *J. ACM (JACM)* **51**, 3, 497–515.
29. Kapitaniak, T. [1994] "Synchronization of chaos using continuous control," *Phys. Rev. E* **50**, 1642–1644.
30. Kiani,–B. A., Fallahi, K., Pariz, N. & Leung, H. [2009] "A chaotic secure communication scheme using fractional chaotic systems based on an extended fractional Kalman filter," *Commun. Nonlinear Sci. Numer. Simulat* **14**, 863–879.
31. Kinzel, W. [2003] "Theory of interacting neural networks," in Bornholdt S. and Schuster H. G. (eds.), Handbook of Graphs and Networks, Wiley-VCH.
32. Kirman, A. [2003] "Economic network," in Bornholdt S. and Schuster H. G. (eds.), Handbook of Graphs and Networks, Wiley-VCH.
33. Kolumban, G., Kennedy, M. P. & Chua, L. O. [1997] "The role of synchronization in digital communications using chaos. Part I: fundamentals of digital communications," *IEEE Trans. Circuits Syst.-I* **44**, 927–936.
34. Kuramoto, Y. [2003] *Chemical oscillations, waves, and turbulence*, Dover Publications.
35. Lampreia, J. P. & Ramos, J. S. [1997] "Symbolic Dynamics for Bimodal Maps," *Portugaliae Math* **54**, 1, 1–18.
36. Miliou, A. N., Valaristos, A. P., Stavrinides, S. G., Kyritsi, K. G. & Anagnostopoulos, A. N. [2007] "Characterization of a non-autonomous second order nonlinear circuit for secure data transmission," *Chaos Solitons & Fractals* **33**, 4, 1248–1255.
37. Milnor, J. & Thurston, W. [1988] "On iterated maps of the interval. Dynamical Systems," *Lecture Notes in Math* Springer, Berlin, **1342**, 465–563.
38. Nagel, K. [2003] "Traffic networks," in Bornholdt S. and Schuster H. G. (eds.), Handbook of Graphs and Networks, Wiley-VCH.
39. Pareek, N. K., Patidar, V. & Sud, K. K. [2005] "Cryptography using multiple one-dimensional chaotic maps," *Commun. Nonlinear Sci. Numer. Simulat* **10**, 715–723.
40. Pecora, L. M. & Carroll, T. L. [1990] "Synchronization in chaotic systems," *Phys. Rev. Lett* **64**, 821–824.
41. Pecora, L. M. & Carroll, T. L. [1991] "Driving systems with chaotic signals," *Phys. Rev. A* **44**, 2374–2383.
42. Pikovsky, A., Rosenblum, M. & Kurths, J. [2001] *Synchronization: a universal concept in nonlinear sciences*, Cambridge University Press.
43. Pyragas, K. [1992] "Continuous control of chaos by self-controlling feedback," *Phys. Lett. A* **170**, 421–428.
44. Rayleigh J. [1945] *The theory of sound*, Dover Publishers, New York.
45. Rocha, J. L. & Ramos, J. S. [2006] "Computing conditionally invariant measures and escape rates," *Neural, Parallel & Sc. Comput* **14**, 97–114.

46. Rulkov, N. F., Sushchik, M. M., Tsimring, L. S. & Abarbanel, H. D. [1995] "Generalized synchronization of chaos in directionally coupled chaotic systems," *Phys. Rev. E* **50**, 1642–1644.
47. Rulkov, N. F. [2001] "Regularization of synchronized chaotic bursts," *Phys. Rev. Lett* **86**, 183–186.
48. Satorras-P, R. & Vespignani, A. [2003] "Epidemics and immunization in scale-free networks," in Bornholdt S. and Schuster H. G. (eds.), Handbook of Graphs and Networks, Wiley-VCH.
49. Soto-Crespo, J. M. & Akhmediev, N. [2005] "Soliton as strange attractor: nonlinear synchronization and chaos," *Phys. Rev. Lett* **95**, 024101-1-4.
50. Stavrinides, S. G., Anagnostopoulos, A. N., Miliou, A. N., Valaristos, A., Magafas, L., Kosmatopoulos, K. & Papaioannou, S. [2009] "Digital chaotic synchronized communication system," *J. Eng. Sci. Tech. Rev* **2**, 82–86.
51. Stojanovski, T., Kocarev, L. & Harris, R. [1997] "Application of symbolic dynamics in chaos synchronization," *IEEE Trans. Circuits and Syst* **44**, 1014–1018.
52. Watts, D. J. & Strogatz, S. H. [1998] "Collective dynamics of small–world networks," *Nature* **393**, 440–442.
53. Strogatz, S. H. [2003] *Sync: The Emerging Science of Spontaneous Order* Hyperion, New York.
54. Tang, M., Liu, L. & Liu, Z. [2009] "Influence of dynamical condensation on epidemic spreading in scale-free networks," *Phys. Rev. E* **79**, 016108-1-6.
55. Teufel, A., Steindl, A. & Troger, H. [2006] "Synchronization of two flow–excited pendula," *Commun. Nonlinear. Sci. Numer. Simulat* **11**, 577–594.
56. Wang, X. Y. & Yu, Q. [2009] "A block encryption algorithm based on dynamic sequences of multiple chaotic systems," *Commun. Nonlinear Sci. Numer. Simulat* **14**, 574–581.
57. Wang, D., Zhong, Y. & Chen, S. [2008] "Lag synchronizing chaotic system based on a single controller," *Commun. Nonlinear Sci. Numer. Simulat* **13**, 637–644.
58. Xiang, T., Wong, K. & Liao, X. [2008] "An improved chaotic cryptosystem with external key," *Commun. Nonlinear Sci. Numer. Simulat* **13**, 1879–1887.
59. Yang, T. & Chua, L. O. [1999] "Generalized synchronization of chaos via linear transformations," *Int. J. Bifurcation & Chaos* **9**, 215–219.
60. Zaks, M. A., Park, E. -H., Rosenblum, M. G. & Kurths, J. [1999] "Alternating locking ratio in imperfect phase synchronization" *Phys. Rev. Lett* **82**, 4228–4231.
61. Zhan, M., Wang, X., Gong, X., Wei, G. W. & Lai C. -H. [2003] "Complete synchronization and generalized synchronization of one way coupled time-delay systems," *Phys. Rev. E* **68** 036208–1–5.

16. [illegible] (199[illegible]) "[illegible] experimental [illegible] of [illegible] in directionally coupled [illegible]," [illegible]
17. [illegible] L. (2007) "[illegible] of synchronized [illegible] bursts," Phys. Rev. Lett. 98, 15[illegible]
18. Sporns, O. [illegible] (2003) "[illegible]" in Bornholdt, S. and Schuster, H. G. (eds.), Handbook of Graphs and Networks (Wiley-VCH, [illegible]).
19. [illegible] (200[illegible]) "Solitons [illegible]," Phys. Rev. E [illegible]
20. Strogatz, S. H., [illegible]
21. [illegible] (200[illegible]) "Application of [illegible]," [illegible]
22. Watts, D. J. & Strogatz, S. H. (1998) "Collective dynamics of 'small-world' networks," Nature 393, 440–442.
23. Strogatz, S. H. (2003) Sync: The Emerging Science of Spontaneous Order (Hyperion, New York).
24. [illegible] (2009) "[illegible]," Phys. Rev. E [illegible]
25. [illegible]
26. Wang, [illegible] (200[illegible]) "[illegible] on the [illegible] systems," [illegible]
27. Wang, [illegible] (2008) "[illegible]," [illegible]
28. [illegible] (199[illegible]) "[illegible]," [illegible] 23, [illegible]
29. Yang, T. & Chua, L. O. (1997) "Generalized synchronization of chaos via linear transformations," Int. J. Bifurcation Chaos 7, 2[illegible]
30. [illegible] (199[illegible]) "[illegible]," Phys. Rev. Lett. [illegible]
31. [illegible] (2008) "[illegible] complex networks [illegible]," [illegible]

# Chapter 8

# Wavelet Study of Dynamical Systems Using Partial Differential Equations

Eugene B. Postnikov

*Department of Theoretical Physics, Kursk State University*
*Radishcheva St., 33, 305000 Kursk, Russia*
*E-mail: postnicov@gmail.com*

The recent development of partial differential equation theory shows that this approach is a powerful tool for the wavelet analysis of non-stationary time series including those generated by chaotic systems. The present paper reviews main directions, where it is applicable, formulates current open problems and possible ways to their solution.

**Keywords**: Wavelets, chaos, computational methods.

## Contents



## 8.1. Definitions and State of Art

The continuous wavelet transform with the Morlet wavelet given by

$$w(a,b) = \int_{-\infty}^{+\infty} f(t) e^{i\omega_0 \frac{t-b}{a}} e^{-\frac{(t-b)^2}{2a^2}} \frac{dt}{\sqrt{2\pi}a} \tag{8.1}$$

is a powerful modern tool for signal and image processing [Mallat, 1999]. Since the kernel in (8.1) is the harmonic function modulated by the Gaussian having the almost-compact support with the characteristic length of scale $a$, this transform allows to evaluate a local spectral analysis, i.e., to find periods $T = 2\pi a/\omega_0$ in the vicinity of time moment $b$.

For example, the perspective objects of its application are the complex solutions of differential equation systems

$$\dot{f}_i(t) = F\left(f_j(t), \dot{f}_j(t)\right)$$

especially the functions that have a chaotic behavior. One of the most important area of applicability is the study of the phenomenon, which is knows as "chaotic synchronization" [Pikovsky *et al.*, 2001]. That means that the absolute value of the instantaneous phase ($\phi_1(t)$ and $\phi_2(t)$) difference of two chaotic functions or time series $f_1(t)$ and $f_2(t)$, must be a bounded function of time: $|\phi_1(t) - \phi_2(t)| < \text{const}$ for any $t$.

However, the definition mentioned above reveals the important problem, which is still far from a complete solution: how to introduce a phase of a chaotic oscillator. The authors of the article [Hramov & Koronovskii, 2004] have mentioned that in some cases there exists a synchrony between phases $\phi(a,b)$ of the complex wavelet transform $w(a,b) = |w(a,b)|\exp(i\phi(a,b))$ defined above (8.1).

Recently, this process has been studied in more details [Postnikov, 2009a]. It is argued that the wavelet phase synchronization provides information about the synchronization of an averaged motion described by bounding tori instead of fine-level classical chaotic phase synchronization. The concepts of bounding tori, first introduced in [Tsankov & Gilmore, 2003] implies that chaotic attractors have various scales of structure: a fine set of unstable orbits at a finer level and a torus with holes, which encloses it. This bounding torus is a semi-penetrable surface defining the domain from which a phase trajectory cannot escape, see the review with applications to various examples including Rössler system, in [Letellier *et al.*, 2006].

The outlooks obtained in [Postnikov, 2009a] are based on the representation (first proposed in [Postnikov, 2007]) of the transform (8.1) in the form $w(\nu,\omega_0,b) = u(\nu,\omega_0,b)\exp(i\pi\nu b)$, where $u(\nu,\omega_0,b)$ admits the diffusion equation

$$\frac{\partial u}{\partial \tau} = \left(2\pi^2\nu^2\right)^{-1}\frac{\partial^2 u}{\partial b^2} \tag{8.2}$$

with the initial condition $u(0,b) = f(b)\exp(-i\pi\nu b)$ at the "instant" $\tau = \omega_0^2$ with the diffusion coefficient $(2\pi^2\nu^2)^{-1}$. Here $\nu = \omega_0/\pi a$ is a parameter corresponding to scale $a$, at which one searches synchronization.

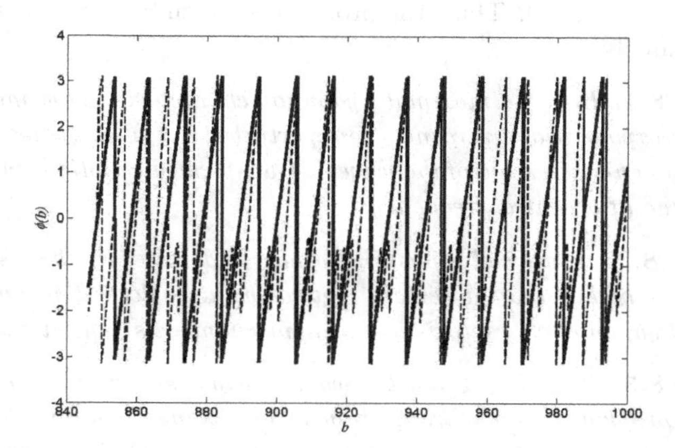


Fig. 8.1. Time dependence of phases for Rössler oscillator: dashed line — original phase angle (that corresponds to the wavelet phase for $\omega_0 = 0$); solid line — the wavelet phase at $\omega_0 = 4\pi$.

In the case if one needs to evaluate the wavelet transform over all possible scales, one can use instead of (8.2) another partial differential equation:

$$\frac{\partial w}{\partial a} = a\frac{\partial^2 w}{\partial b^2} - i\omega_0 \frac{\partial w}{\partial b} \tag{8.3}$$

It has been proven in [Postnikov, 2006] that the transform (8.1) is a solution of the Cauchy problem for (8.3) with the initial condition $w(0, b) = f(b)\exp(-\omega^2/2)$.

The detailed derivation of PDEs (8.2), (8.3), discussion of their applicability and computational realization, as well as an illustrated by examples comparison with other wavelet methods can be found in the comprehensive review [Postnikov, 2009b].

## 8.2. Open Problems in the Continuous Wavelet Transform and a Topology of Bounding Tori

Recently, it has been shown in [Postnikov, 2009a] that the so-called "wavelet phase" or "time-scale"synchronization of chaotic signals is actually synchronization of smoothed functions with reduced chaotic fluctuations, see Fig. 8.1. One can see that the original phase curve considered as a polar angle on a phase plane is not a pure saw-like line but has random bursts due to chaoticity. This bursts are the chaotic fluctuations eliminated by diffusion process (8.2). Thus, the problem is to study this process in more details. Namely:

**Problem 8.1.** *How the maximal Lyapunov characteristic exponent, giving the rate of exponential divergence from perturbed initial conditions, and calculated along a given scale of the wavelet-transformed chaotic signal depends on the value of a central frequency.*

**Problem 8.2.** *A wavelet synchronization is detected if the used central frequency is higher than a certain critical value – does it means that the maximal Lyapunov characteristic exponent change its sign at this value?*

**Problem 8.3.** *It is argued that the wavelet phase synchronization provides information about the synchronization of an averaged motion described by bounding tori- is it possible to determine the radii of these tori using the data from the wavelet transform?*

## 8.3. The Evaluation of the Continuous Wavelet Transform Using Partial Differential Equations in Non-Cartesian Co-ordinates and Multidimensional Case

The known [Postnikov, 2006] representation of the continuous Morlet wavelet transform as a solution of the Cauchy problem for a complex partial differential equation (8.3) is valid for 1D Cartesian co-ordinates only.

Correspondingly, the following generalizations are desirable:

**Problem 8.4.** *Is it possible to use a similar PDE approach to have the complex wavelet transform for a local spectral analysis in the case of non-Cartesian or, in general, arbitrary curvilinear coordinate systems? It is natural that the corresponding wavelet will consists another orthogonal function instead of complex exponential (say, Bessel functions, in the case of cylindrical co-ordinates), but will this wavelet generated by simple replacing of derivatives in (8.3) by their curvilinear analogs.*

**Problem 8.5.** *To generalize Eq. (8.3) on 2D case and to use it for 2D spectral image processing; to study problems connected with anysotropy of basis wavelets in two-dimensional case.*

**Problem 8.6.** *To generalize Eq. (8.3) on an arbitrarily many-dimensional case, including partial differential equations on manifolds and to analysis an existence of PDE-generated wavelet transform in these cases, as well as their applicability for study of multidimensional dynamic systems.*

## 8.4. Discussion of Open Problems

The open problems 1 and 2 are connected with the calculation of a basic test of chaos presence: the value of largest Lyapunov characteristic exponent (LCE). A dynamical system is chaotic if at least one LCE is positive. Thus, the idea is the following: if one replaces the original phase curve, having positive LCE, with the phase curve obtained by the wavelet transform at a given scale, i.e., $(x(t), y(t)) \to (\text{Re}w(b,a), \text{Im}w(b,a))$ for some fixed $a$, Here the complexification $f(t) = x(t) + iy(t)$ can be used as the transformed function in (8.1) and, correspondingly, in the initial condition for (8.2). The main challenge is to prove the supposition that the wavelet synchronization is strictly connected with positivity of LCE in the wavelet space.

The second set of problems could be solved by replacing of the simple 1D Cartesian form of derivatives in (8.2) by their co-ordinate-free vector version

$$\frac{\partial w}{\partial a} = a\nabla^2 w + \vec{\Omega} \cdot \nabla w \tag{8.4}$$

This is a natural way to the definition of the directional Morlet wavelet. Namely, if $\Omega = (\omega_x, \omega_y)$, then $\alpha = \text{atan}(\omega_y/\omega_x)$ is the angle determining the direction of the wavelet. Additionally, the form (8.4) provides an opportunity to apply the method in the case of non-Cartesian grids, say on a sphere (in this case one needs only to represent the Laplacian and the gradient operators in the spherical co-ordinates) or on manifolds.

## References


1. Hramov, A. E. & Koronovskii, A. A. [2004] "Wavelet transform analysis of the chaotic synchronization of dynamical systems," *JETP Lett* **79**, 316–319.
2. Letellier, C., Roulin, E. & Rössler, O. E. [2006] "Inequivalent topologies of chaos in simple equations," *Chaos, Solitons & Fractals* **28**, 337–360.
3. Mallat, S. [1999] *A Wavelet Tour of Signal Processing*, Academic Press.
4. Pikovsky, A. Rosenblum, M. & Kurths, J. [2001] *Synchronization: An Universal Concept in Nonlinear Sciences*, Cambridge University Press.
5. Postnikov, E. B. [2007] "On precision of wavelet phase synchronization of chaotic systems," *J. Exp. Theor. Phys* **105**, 652–654.
6. Postnikov, E. B. [2007] "Evaluation of a continuous wavelet transform by solving the Cauchy problem for a system of partial differential equations," *Comp. Math. Math. Phys* **46**, 73–78.
7. Postnikov, E. B. [2009a] "Wavelet phase synchronization and chaoticity," *Phys. Rev. E* **80**, 057201.

8. Postnikov, E. B. [2009b] "Partial differential equations as a tool for evaluation of the continuous wavelet transform," in *Mathematical Physics Research Developments* ed. M. B. Levy, Nova Science Publishers, 1–36.
9. Tsankov, T. D & Gilmore, R. [2003] "Strange attractors are classified by bounding tori," *Phys. Rev. Lett* **91**, 134104.

# Chapter 9

# Combining the Dynamics of Discrete Dynamical Systems

Jose S. Cánovas*

*Departamento de Matemática Aplicada y Estadística*
*Universidad Politécnica de Cartagena*
*C/ Dr. Fleming sn, 30202 Cartagena, Murcia, Spain*
*E-mail: Jose.Canovas@upct.es*

We introduce several problems involving non-autonomous discrete systems which are generated by periodic sequences of continuous maps.

**Keywords**: Combining dynamics, discrete dynamical systems, Sharkovsky order, commutativity problems, triangular maps on the square.

## Contents



## 9.1. Introduction

Let $(X, d)$ be a compact metric space and let $f_{1,\infty} = (f_n)_{n=1}^{\infty}$ be a sequence of continuous maps $f_n : X \to X$. A non-autonomous discrete system is the pair $(X, f_{1,\infty})$. Given $x \in X$, its orbit is the sequence $(f_1^n(x))_{n=0}^{\infty}$, where

*The author has been partially supported by the grants MCI (Ministerio de Ciencia e Innovación) and FEDER (Fondo Europeo de Desarrollo Regional), grant MTM2008-03679/MTM and 08667/PI/08 (Fundación Séneca, CARM).

$f_1^0$ is the identity on $X$, $f_1^1 = f_1$ and $f_1^n = f_n \circ ... \circ f_2 \circ f_1$, for $n \geq 2$. Let $\text{Orb}_{f_{1,\infty}}(x)$ denote the orbit of $x$ by $f_{1,\infty}$.

This work is devoted to introduce some open problems concerning periodic non-autonomous discrete systems, that is, when there is a minimal integer $T$ such that $f_{n+T} = f_n$ for all $n \in \mathbb{N}$. Note that in this case, $f_1^1 = f_1$ and $f_1^n = f_{(n)} \circ f_{(n-1)} \circ ... \circ f_1$, where $(n) \in \{1, ..., T\}$ holds that $n - (n)$ is divided by $T$. Hence, the non-autonomous system can be thought as the combination of the maps $f_1$,...,$f_T$.

When $f_{1,\infty} = (f, f, f, ...)$, that is $T = 1$, we have a discrete dynamical system, which will be denoted by replacing $f_{1,\infty}$ by $f$, as $(X, f)$. We are going to introduce some definitions for non-autonomous systems $(X, f_{1,\infty})$ which have an immediate translation to discrete dynamical systems. In that case we will write $f$ instead of $f_{1,\infty}$, without introducing the analogous definition.

The motivations for studying these systems comes from several applied sciences. For instance, let us introduce the following model which comes from economic dynamics (see [Puu, 1991]). Assume a market consisting in two firms producing equivalent goods, with isoelastic demand function (in its inverse form)

$$p = \frac{1}{q_1 + q_2} \tag{9.1}$$

where $q_i$, $i = 1, 2$ are the outputs of both firms and $p$ is the price. In addition, assume that the cost functions are given by $C_i = c_i \cdot q_i$, $i = 1, 2$, being the constant marginal costs $c_i$. Each firm wants to maximize the profit function

$$\Pi_i = p \cdot q_i - c_i \cdot q_i = \frac{q_i}{q_1 + q_2} - c_i \cdot q_i \tag{9.2}$$

Solving the equation $\frac{\partial \Pi_i}{\partial q_i} = 0$, we obtain that

$$q_1 = \sqrt{q_2/c_1} - q_2 = f_1(q_2) \tag{9.3}$$

and

$$q_2 = \sqrt{q_1/c_2} - q_1 = f_2(q_1) \tag{9.4}$$

The maps $f_1$ and $f_2$ are called reaction functions, because firms plan their future production according to these maps (note that firms' aim is to maximize their profits). The reaction functions have a one dimensional domain and the productions of each firm are given by the periodic sequences $(f_1, f_2, f_1, f_2, ...)$ and $(f_2, f_1, f_2, f_1, ...)$, respectively. We refer the reader to the references [Dana & Montrucchio, 1986, Kopel, 1996, Puu & Norin, 2003] for additional examples of these kind of duopolies.

Periodic models can also appear because of model time dependance. For instance, economic dynamics gives us examples of markets with $n$ firms (oligopolies) and investment periods. In that investment periods, the firms are forced to invest part of their capital stock. Hence, we have two different reaction functions, one for the investment period and another different one when the firms do not invest. In the simplest case of periodic investment periods, these oligopoly models are given by periodic sequences of maps (see e.g. [Puu, 2005]).

Let us emphasize that recently, some scientists working on population dynamics use such kind of systems to model the population growth of species under some periodic changes in the environment. For instance, the non-autonomous Beverton-Holt equation

$$x_{n+1} = \frac{\mu K_n x_n}{K_n + (\mu - 1)x_n} \tag{9.5}$$

where $K_n$ is a periodic sequence. This non-autonomous equation has been studied in [Cushing & Henson, 2002; Elaydi & Sacker, 2005]. The sequence $K_n$ is the "carrying capacity of the environment" and $\mu$ is the "inherent growth rate" of the population $x_n$.

Finally, we must remark the existence of the so-called Parrondo's paradox [Harmer & Abbott, 1999], which has been studied in many applied sciences like physics [Parrondo *et al.*, 2000], economy [Spurgin & Tamarkin, 2005] or biomathematics [Steele, 1998]. Roughly speaking, it consists in alternating different games (in a stochastic or deterministic way) and comparing certain properties of the combined game with the properties of the individual games. There exists paradox if these properties are completely different. Parrondo's paradox has been extended to the context of nonlinear dynamical systems. In this case, the most relevant property to study is the chaotic behavior of the single and combined dynamics. For instance, let $(X, f_i)$, $i = 1, 2$, be two discrete dynamical systems and consider the combined system $(X, f_{1,\infty})$ where $f_{1,\infty}$ is the periodic sequence $(f_1, f_2, f_1, f_2, ...)$. There is paradox when the dynamical behaviors of $(X, f_i)$ are simple (resp. complicated) and $(X, f_{1,\infty})$ is complicated (resp. simple). For instance, in [Almeida, *et al.*, 2005] we can see that the phenomenon "chaos+chaos=order" is possible when periodic combinations of one-dimensional quadratic maps are taken. In the periodic case, the dynamic Parrondo's paradox was deeply analyzed in [Cánovas *et al.*, 2006] (below we will explain the obtained results and the related open problems). For the sake of completeness, let us emphasize that the dynamic Parrondo's paradox for random combination of maps was analyzed by Boyarsky and collaborators, showing the existence of absolutely continuous measures in the case of random combinations of simple piecewise smooth maps [Boyarsky *et al.*, 2005].

The work is organized as follows. The next section will be devoted to introduce some basic notation while the open problems will be stated in Sec. 9.3.

## 9.2. Basic Definitions and Notations

Let $(X, d)$ be a compact metric space and denote by $C(X, X)$ the set of continuous maps $f : X \to X$. Fix $(X, f_{1,\infty})$ a non-autonomous discrete system. As we mentioned above, all the definitions and notation on non-autonomous systems have an immediate translation to dynamical systems, just considering $f_{1,\infty}$ as a constant sequence $(f, f, f, ...)$, and then writing $f$ instead of $f_{1,\infty}$.

An orbit $\mathrm{Orb}_{f_{1,\infty}}(x) = (x_n)_{n=1}^{\infty}$ is periodic if there is $k \in \mathbb{N}$ such that $x_n = x_{n+k}$ for all $n \in \mathbb{N}$. The smallest integer $k$ satisfying the above condition is called the period of $x = x_1$, which is called a periodic point of $f_{1,\infty}$. Periodic points of period one are called fixed points of $f_{1,\infty}$. We denote by $\mathrm{Per}(f_{1,\infty})$ the set of periods of $f_{1,\infty}$, and by $\mathrm{Fix}(f_{1,\infty})$ and $\mathrm{P}(f_{1,\infty})$ the sets of fixed and periodic points of $f_{1,\infty}$, respectively. Note that $x \in \mathrm{Fix}(f_{1,\infty})$ if and only if $x \in \mathrm{Fix}(f_n)$ for all $n \in \mathbb{N}$.

Consider the Sharkovsky order in the set of natural numbers "$\succ_{s}$", given by

$$2^n \cdot p \succ_{s} 2^m \cdot q \tag{9.6}$$

for $n, m \in \mathbb{N} \cup \{0\}$ and $p, q$ odd numbers, if one of the following conditions is fulfilled

- $p = q = 1$ and $n > m$,
- $q = 1$ and $p > 1$,
- $1 \notin \{p, q\}$ and $n < m$,
- $1 < p < q$ and $n = m$.

For instance, $3 \succ_{s} n$ for all $n \in \mathbb{N}$, $4 \succ_{s} 2 \succ_{s} 1$ and $7 \succ_{s} 10 \succ_{s} 20 \succ_{s} 16$.

For $n \in \mathbb{N} \cup \{2^\infty\}$, let

$$\mathcal{S}(n) := \{m \in \mathbb{N} : n \succ_{s} m\} \cup \{n\} \tag{9.7}$$

and

$$\mathcal{S}(2^\infty) := \{2^n : n \in \mathbb{N} \cup \{0\}\} \tag{9.8}$$

Sharkovsky's Theorem states the following (see [Sharkovsky, 1964] and for a simple proof, [Du, 2004]).

**Theorem 9.1.** *Let $f \in C(I, I)$, $I = [0, 1]$, be such that it has a periodic orbit (periodic sequence) of period $n$. Then $f$ has periodic points (periodic sequences) of period $m \in S(n)$. Moreover, for any $n \in \mathbb{N} \cup \{2^\infty\}$ there is $f_n \in C(I)$ such that* $\mathrm{Per}(f_n) = \mathcal{S}(n)$.

Fixed and periodic points are the strongest notions of recurrence in dynamical systems. There are weaker notions of recurrence that contain the sets of both fixed and periodic points (see e.g. [Block & Coppel, 1992]). A point $x \in X$ is called recurrent if for any open neighborhood $U$ of $x$ there is a strictly increasing sequence $(n_i)_{i=1}^\infty$ such that $f^{n_i}(x) \in U$. If the sequence $(n_i)_{i=1}^\infty$ has bounded gaps, the point is called uniformly recurrent (also called regularly recurrent). If $n_i = ki$ for some $k \in \mathbb{N}$ the point is called almost periodic. Denote by $\mathrm{Rec}(f)$, $\mathrm{UR}(f)$ and $\mathrm{AP}(f)$ the sets of recurrent, uniformly recurrent and almost periodic points, respectively. It is clear from the definitions that

$$\mathrm{Fix}(f) \subseteq \mathrm{P}(f) \subseteq \mathrm{AP}(f) \subseteq \mathrm{UR}(f) \subseteq \mathrm{Rec}(f) \tag{9.9}$$

For any $x \in X$, we denote by $\omega_{f_{1,\infty}}(x)$ the set of limit points of the orbit $\mathrm{Orb}_{f_{1,\infty}}(x)$, that is, $y \in \omega_{f_{1,\infty}}(x)$ if there is a strictly increasing sequence of non-negative integers $(n_i)_{i=0}^\infty$ such that

$$\lim_{i \to \infty} f_1^{n_i}(x) = y \tag{9.10}$$

When $f_{1,\infty} = f$, the set $\omega_f(x)$ is compact and invariant by $f$, that is, $f(\omega_f(x)) \subseteq \omega_f(x)$. Note that $x \in \mathrm{Rec}(f)$ if and only if $x \in \omega_f(x)$. On the other hand, for any $x \in \mathrm{UR}(f)$, the $\omega$-limit set $\omega_f(x)$ is a minimal set, that is, it does not contain any proper closed subset which is invariant by $f$.

When there exists $x \in X$ such that $\omega_{f_{1,\infty}}(x) = X$, we will say that $f_{1,\infty}$ is transitive. The transitivity of an interval map $f \in C(I, I)$ is equivalent to chaoticity in Devaney's sense. Recall that $f \in C(X, X)$ is chaotic in Devaney's sense (see [Devaney, 1989]) if it satisfies the following conditions:

- $f$ is transitive.
- $\mathrm{P}(f)$ is dense in $X$.

- $f$ has sensitive dependence on initial conditions, that is, there is $\varepsilon > 0$ such that for any $x \in I$ there is an arbitrarily close $y \in I$ and $n \in \mathbb{N}$ such that $d(f^n(x), f^n(y)) > \varepsilon$.

One of the main topics in dynamical systems is the dynamical complexity or chaos. A well-known measure of such complexity is given by topological entropy (see [Kolyada & Snoha, 1996]). Given $\epsilon > 0$ and $n \in \mathbb{N}$, we say that a subset $E \subset X$ is $(n, \epsilon, f_{1,\infty})$-separated if for any $x, y \in E$ there is $k \in \{0, 1, ..., n-1\}$ such that $d(f_1^k(x), f_1^k(y)) > \epsilon$. Denote by $s(n, \epsilon, f_{1,\infty})$ the cardinality of a maximal $(n, \epsilon, f_{1,\infty})$-separated subset of $X$. The topological entropy of $f_{1,\infty}$ is the number defined by:

$$h(f_{1,\infty}) = \lim_{\epsilon \to 0} \limsup_{n \to \infty} \frac{1}{n} \log s(n, \epsilon, f_{1,\infty}) \tag{9.11}$$

In the autonomous case, when $f_{1,\infty}$ is a constant sequence of maps, topological entropy is taken as a measure of the dynamic complexity of the discrete system (see [Adler *et al.*, 1965] for the seminal definition, and [Block & Coppel, 1992] and [Alsedá *et al.*, 1993, Chapter 4] for classical books).

Useful properties of topological entropy of a continuous map $f \in C(X, X)$ are the following (see e.g. [Alsedá *et al.*, 1993, Chapter 4]):

- **Conjugacy invariancy**. Let $f : X \to X$ and $g : Y \to Y$ be continuous maps on compact metric spaces and let $\pi : X \to Y$ be a continuous surjective map satisfying $\pi \circ f = g \circ \pi$. Then $h(g) \leq h(f)$. If in addition $\pi$ is an homeomorphism, $g$ and $f$ are said to be conjugated and $h(g) = h(f)$.
- If $X = \cup_{i=1}^k X_k$, where $X_k \subset X$ are invariant by $f$ compact subsets, then

$$h(f) = \max\{h(f|_{X_i}) : 1 \leq i \leq k\} \tag{9.12}$$

- If $X = I = [0, 1]$ and $f \in C(I, I)$ is piecewise monotone, that is, there are $0 = x_0 < x_1 < ... < x_{n-1} < x_n = 1$ such that $f|_{[x_i, x_{i+1}]}$ is monotone for $i = 0, 1, ..., n-1$, then

$$h(f) = \lim_{n \to \infty} \frac{1}{n} \log c(f^n) \tag{9.13}$$

where $c(f^n)$ is the number of monotone pieces of $f^n$ (see [Misiurewicz & Szlenk, 1980]).

It is remarkable that if $h(f)$ is positive, then the map $f$ is chaotic in the sense of Li and Yorke (see [Blanchard *et al.*, 2002]). In general, a sequence

$f_{1,\infty}$ is chaotic in the sense of Li and Yorke if there is an uncountable subset $S \subset X$ such that for any $x, y \in S$, $x \neq y$, we have that

$$\liminf_{n\to\infty} d(f_1^n(x), f_1^n(y)) = 0 \tag{9.14}$$

and

$$\limsup_{n\to\infty} d(f_1^n(x), f_1^n(y)) > 0 \tag{9.15}$$

So, $S$ is called an scrambled set for $f_{1,\infty}$.

In the case of continuous maps on the interval, there is a clear dichotomy between Li-Yorke chaos and simplicity. Namely, we say that an orbit $\mathrm{Orb}_f(x)$ is approximated by periodic orbits if for any $\varepsilon > 0$ there is $y \in \mathrm{P}(f)$ and $n_0 \in \mathbb{N}$ such that $|f^n(x) - f^n(y)| < \varepsilon$. So, the map $f$ is either Li-Yorke chaotic or any orbit is approximated by periodic orbits [Smital, 1986]. Below, we will introduce our problems.

## 9.3. Statement of the Problems

As a first approach, let us remark that periodic non-autonomous systems can be studied as a dynamical system. Given $x \in X$, its orbit under $f_{1,\infty}$ can be analyzed by means of some of its subsequences. More precisely, if $\mathrm{Orb}_{f_{1,\infty}}(x) = (x_n)_{n=0}^{\infty}$, then, for $i = 0, 1, ..., T-1$, the subsequences $(x_{n \cdot T+i})_{n=0}^{\infty}$ are exactly $\mathrm{Orb}_{f_i \circ f_{i-1} \circ ... \circ f_1 \circ f_T \circ f_{T-1} \circ ... \circ f_{i+1}}(f_1^i(x))$. So, for any $x \in X$, it is simple to prove that

$$\omega_{f_{1,\infty}}(x) = \omega_{f_T \circ ... \circ f_1}(x) \cup \omega_{f_1 \circ f_T \circ ... \circ f_2}(f_1(x)) \cup ... \cup \omega_{f_{T-1} \circ ... f_1 \circ f_T}(f_1^{T-1}(x)) \tag{9.16}$$

Hence, one can wonder whether the behavior of $f_{1,\infty}$ can be deduced from the behavior of the compositions $f_T \circ ... \circ f_1$, $f_1 \circ f_T \circ ... \circ f_2$, ... and $f_{T-1} \circ ... f_1 \circ f_T$.

This idea produces another positive results, as for instance, in computing the topological entropy. When $f_{1,\infty}$ is a periodic sequence, it is proved in [Kolyada & Snoha, 1996] that

$$h(f_{1,\infty}) = h(f_T \circ ... \circ f_1) = h(f_1 \circ f_T \circ ... \circ f_2) = ... = h(f_{T-1} \circ ... f_1 \circ f_T) \tag{9.17}$$

Finally, in [AlSharawi *et al.*, 2006] a characterization of periodic solutions of one dimensional non-autonomous difference equations has been found in terms of the Sharkovsky's result for one dimensional maps. Namely, for $q \in \mathbb{N}$, let

$$\mathcal{A}_q = \{n : \mathrm{lcm}(n, T) = q \cdot T\} \tag{9.18}$$

Notice that $p \cdot q \in \mathcal{A}_q$ and define the equivalence relationship "$\sim$" on $\mathbb{N}$ by stating that $n \sim m$, $n, m \in \mathbb{N}$, if and only if $n$ and $m$ belong to the same set $\mathcal{A}_q$. If we denote any equivalence class $\mathcal{A}_q$ by $[q]$, we define the order on $\mathbb{N}/\sim$ by $[n] >_s [m]$ if and only if $n >_s m$. Then it is proved that if $[n] \in \mathrm{Per}(f_{1,\infty})/\sim$, then any $[m] \in \mathbb{N}/\sim$ such that $[n] >_s [m]$, then $[m] \in \mathrm{Per}(f_{1,\infty})/\sim$. The proof of this result is based on two facts: Sharkovsky's Theorem and the fact that if $m \in \mathrm{Per}(f_{1,\infty})$ and $m \in [q]$, then $q \in \mathrm{Per}(f_T \circ ... \circ f_1)$.

However, we must point out that the above compositions, $f_1 \circ f_2$ and $f_2 \circ f_1$, may need not have the same dynamic properties, as we will show in the section devoted to commutativity problems (see [Balibrea, *et al.*, 1999] or [Cánovas *et al.*, 2006]). Additionally, some dynamic properties of a sequence $f_{1,\infty}$ cannot be obtained from the above compositions; for instance, the existence of periodic orbits of odd period of a non-autonomous system $f_{1,\infty} = (f_1, f_2, f_1, f_2, ...)$ defined on the unit interval $[0, 1]$ cannot be deduced from the periodic orbits of $f_1 \circ f_2$ and $f_2 \circ f_1$ (see [Cánovas & Linero, 2006]). In addition, if we define

$$\mathbb{N}^* = \mathbb{N} \setminus (\{2n - 1 : n \in \mathbb{N}\} \cup \{2\}) \tag{9.19}$$

it is proved the following result.

**Theorem 9.2.** *Let* $f_1, f_2 \in C(I)$. *Then*

(a) *If* $f_{1,\infty}$ *has a periodic orbit of period* $n \in \mathbb{N}^* \cup \{2^\infty\}$,

$$\mathcal{S}(n)\backslash\{1, 2\} \subset \mathrm{Per}(f_{1,\infty}).$$

(b) *If* $2n + 1 \in \mathrm{Per}(f_{1,\infty})$, $n \geq 1$,

$$\mathcal{S}(2 \cdot 3)\backslash\{1\} \subset \mathrm{Per}(f_{1,\infty}).$$

(c) *There are* $f_1, f_2 \in C(I)$ *such that* $\mathrm{Per}(f_{1,\infty})$ *is* $\{1\}$, $\{2\}$ *or* $\{1, 2\}$.

(d) *For any* $n \in \mathbb{N}^* \cup \{2^\infty\}$:

  *d.1. There are* $f_1, f_2 \in C(I)$ *such that* $\mathrm{Per}(f_{1,\infty}) = \mathcal{S}(n)$.

  *d.2. There are* $f_1, f_2 \in C(I)$ *such that* $\mathrm{Per}(f_{1,\infty}) = \mathcal{S}(n)\backslash\{1\}$.

  *d.3. There are* $f_1, f_2 \in C(I)$ *such that* $\mathrm{Per}([f_{1,\infty}) = \mathcal{S}(n)\backslash\{2\}$.

(e) *Let* $\mathrm{Imp} \subseteq \{2n + 1 : n \in \mathbb{N}\}$. *Then*

  *e.1. For any subset of odd numbers* Imp *there are* $f_1, f_2 \in C(I)$ *such that* $\mathrm{Per}(f_{1,\infty}) = \mathrm{Imp} \cup (\mathcal{S}(2 \cdot 3)\backslash\{1\})$.

  *e.2. For any subset of odd numbers* Imp *there are* $f_1, f_2 \in C(I)$ *such that* $\mathrm{Per}(f_{1,\infty}) = \mathrm{Imp} \cup \mathcal{S}(2 \cdot 3)$.

Notice that the case $\text{Per}(f_{1,\infty}) = \text{Imp}\cup(\mathcal{S}(2\cdot3)\backslash\{2\})$ is not allowed, that is, if $2n+1 \in \text{Per}(f_{1,\infty})$ for some $n \in \mathbb{N}$, then automatically $2 \in \text{Per}(f_{1,\infty})$. In addition, for $n \in \mathbb{N}^* \cup \{2^\infty\}$, $n \neq 2 \cdot 3$, there are not continuous maps $f_1$ and $f_2$ such that $\text{Per}(f_{1,\infty}) = \text{Imp} \cup (\mathcal{S}(n)\backslash\{1\})$ or $\text{Per}(f_{1,\infty}) = \text{Imp} \cup (\mathcal{S}(n)\backslash\{2\})$ or $\text{Per}(f_{1,\infty}) = \text{Imp} \cup \mathcal{S}(n)$. The key for proving Theorem 9.2 is that, given $x \in [0,1]$, its orbit $\text{Orb}_{f_{1,\infty}}(x)$ is periodic with odd period $q$ if and only if

$$\text{Orb}_{f_{1,\infty}}(x) = \text{Orb}_{f_1}(x) = \text{Orb}_{f_2}(x) \tag{9.20}$$

and this property cannot be deduced from the composition $f_2 \circ f_1$. We can identify three types of different problems with periodic non-autonomous systems:

- Can a dynamical property of $f_{1,\infty}$ be studied from the analogous dynamical properties of $f_T \circ ... \circ f_1$, $f_1 \circ f_T \circ ... \circ f_2, ...$ and $f_{T-1} \circ ... f_1 \circ f_T$?
- Are the dynamical properties of $f_T \circ ... \circ f_1$, $f_1 \circ f_T \circ ... \circ f_2, ...$ and $f_{T-1} \circ ... f_1 \circ f_T$ equivalent?. That is, if $f_T \circ ... \circ f_1$ has a dynamical property the maps $f_1 \circ f_T \circ ... \circ f_2, ...$ and $f_{T-1} \circ ... f_1 \circ f_T$ have the same property?
- Can a non-autonomous periodic system $(X, f_{1,\infty})$ be studied from its individual systems $(X, f_i)$, $i = 1, ..., T$?

Below, we state concrete problems concerning these general problems.

### 9.3.1. *Dynamic Parrondo's Paradox and Commuting Functions*

Denote by $\mathcal{D}(f_{1,\infty})$ the set of dynamic properties of $f_{1,\infty}$. Dynamic properties of $f_{1,\infty}$ are, for instance, positive topological entropy, existence of periodic orbits of period $m \in \mathbb{N}$, or to exhibit chaos in the sense of Li and Yorke.

Fix $X = [0,1]$ and a continuous map $f : [0,1] \to [0,1]$. Let $J = [a,b] \subseteq [0,1]$ and denote by $\varphi_J : J \to [0,1]$ a linear map such that $\varphi_J(a) = 0$ and $\varphi_J(b) = 1$. Define $f_J : [0,1] \to [0,1]$ by $f_J(x) = \varphi_J^{-1} \circ f \circ \varphi_J(x)$ if $x \in J$, $f_J(0) = 0$, $f_J(1) = 1$, and linear on any connected component of $[0,1] \setminus J$. A dynamic property $P \in \mathcal{D}(f)$ is an L-property if for any continuous map $f$ and any compact subinterval $J \subseteq [0,1]$, it is held that $P \in \mathcal{D}(f) \cap \mathcal{D}(f_J)$.

Since $h(f) = h(f_J)$, the topological entropy is an L-property. Additional examples of L-properties are to be chaotic in the sense of Li-Yorke or $\text{Per}(f) = \mathcal{S}(n)$, $n \in \mathbb{N}\cup\{2^\infty\}$. However, there are dynamic properties which are not L-properties as for instance the existence of dense orbits (topological transitivity). The next result from [Cánovas *et al.*, 2006] is very useful to realize that the construction of continuous interval maps which exhibit the dynamic Parrondo's paradox is relatively easy.

**Theorem 9.3.** *Let $P_i$, $i = 1, 2, 3$, be L–properties. Then there are continuous maps $f$ and $g$ such that $P_1 \in \mathcal{D}(f)$, $P_2 \in \mathcal{D}(g)$ and $P_3 \in \mathcal{D}(g \circ f)$.*

The construction of maps for proving Theorem 9.3 can be done as follows. Let $f_i \in C(I, I)$ be such that $P_i \in \mathcal{D}(f_i)$, $i = 1, 2, 3$. Since $P_i$ are L-properties, we may assume that $f_i(0) = 0$ and $f_i(1) = 1$ for $i = 1, 2, 3$. Let $J_i = [(i-1)/3, i/3]$, $i = 1, 2, 3$. Let $\psi(x) = 1 - x$, $x \in I$. Define the maps

$$f(x) = \begin{cases} 1 & \text{if } x \in J_1 \\ (\varphi_{J_3}^{-1} \circ \psi \circ f_3 \circ \varphi_{J_2})(x) & \text{if } x \in J_2 \\ (\varphi_{J_3}^{-1} \circ f_1 \circ \varphi_{J_3})(x) & \text{if } x \in J_3 \end{cases} \tag{9.21}$$

and

$$g(x) = \begin{cases} (\varphi_{J_1}^{-1} \circ f_2 \circ \varphi_{J_1})(x) & \text{if } x \in J_1 \\ x & \text{if } x \in J_2 \\ (\varphi_{J_2}^{-1} \circ \psi \circ \varphi_{J_3})(x) & \text{if } x \in J_3 \end{cases} \tag{9.22}$$

Notice that any orbit of $f$ is eventually in $J_3$ and any orbit of $g$ is eventually in $J_1$, or it is a fixed point, and hence $f$ and $g$ have the same properties that $f_1$ and $f_2$. Then $P_1 \in \mathcal{D}(f)$ and $P_2 \in \mathcal{D}(g)$. Notice also that $P_3 \notin \mathcal{D}(f)$ (resp. $P_3 \notin \mathcal{D}(g)$) unless $P_3 \in \mathcal{D}(f_1)$ (resp. $P_3 \notin \mathcal{D}(f_2)$). A simple computation gives

$$(g \circ f)(x) = \begin{cases} 1/3 & \text{if } x \in J_1 \\ (\varphi_{J_2}^{-1} \circ f_3 \circ \varphi_{J_2})(x) & \text{if } x \in J_2 \\ (\varphi_{J_2}^{-1} \circ \psi \circ f_1 \circ \varphi_{J_3})(x) & \text{if } x \in J_3 \end{cases} \tag{9.23}$$

and then $P_3 \in \mathcal{D}(g \circ f)$. Notice again that by the construction $P_i \notin \mathcal{D}(g \circ f)$ unless $P_i \in \mathcal{D}(f_3)$, $i = 1, 2$.

In particular, we can construct maps such that $f$ and $g$ have a complicated (simple) L-property and $f \circ g$ has not this property [$P \in \mathcal{D}(f) \cap \mathcal{D}(g)$ and $P \notin \mathcal{D}(g \circ f)$]. For instance, we consider the topological entropy. From Theorem 9.3, we can construct two continuous interval maps, $f$ and $g$, with zero topological entropy (and hence simple) such that $f \circ g$ has positive topological entropy (and therefore a complicated dynamics). Let us remark that the opposite result is possible.

The construction of the maps $f$ and $g$ in the proof of Theorem 9.3 do not commute. So, our first open problem is related to add the commutativity condition to the hypothesis of Theorem 9.3 as follows.

**Problem 9.1.** *Is it possible to prove Theorem 9.3 with the additional property that the maps $f$ and $g$ commute?*

It is simple to construct two commuting continuous maps $f$ and $g$ with L-properties $P_1$ and $P_2$, respectively, and such that the composition $f \circ g = g \circ f$ is constant. Just consider the subintervals $I_1 = [0, 1/2]$ and $I_2 = [1/2, 1]$, and two continuous interval maps $f_1, g_1 : [0,1] \to [0,1]$ with the desired L-properties $P_1$ and $P_2$ and such that $f_1(1) = f_2(0) = 0$. Then, define the maps

$$f(x) = \begin{cases} \varphi_{I_1}^{-1} \circ f_1 \circ \varphi_{I_1}(x) & \text{if } x \in I_1 \\ \frac{1}{2} & \text{if } x \in I_2 \end{cases} \tag{9.24}$$

and

$$g(x) = \begin{cases} \frac{1}{2} & \text{if } x \in I_1 \\ \varphi_{I_2}^{-1} \circ g_1 \circ \varphi_{I_2}(x) & \text{if } x \in I_2 \end{cases} \tag{9.25}$$

to finish the construction. Then $(f \circ g)(x) = (g \circ f)(x) = 1/2$ for all $x \in I$. Constructing commuting maps such that $f \circ g$ has additional L-properties is a more complicated question.

On the other hand, the maps in the proof of Theorem 9.3 are constructed on subintervals piece by piece. Often, models are given by an analytical expression, as for instance the logistic family $f(x) = ax(1-x)$, $a \in [1, 4]$. So, studying the Parrondo's paradox in these "natural" families is an immediate question.

**Problem 9.2.** *Let $f_a : I \to I$, $I = [0, 1]$, be a one parameter family of maps. For $a_1$ and $a_2$ in the parameter family, find conditions on the parameters such that $f_{a_1}$ and $f_{a_2}$ hold the dynamic Parrondo's paradox.*

As a first approach, one could consider the very well-known studied logistic family $f_a(x) = ax(1-x)$, $x \in I$ and $a \in [1, 4]$. Some numerical experiments showing the existence of paradox for a one parameter family of maps can be found in [Almeida *et al.*, 2005]. However, we claim for the possibility of proving the existence of paradox by means of analytical proofs. In this setting, we point out the works [Andrecut & Ali, 2001(a-b)] which show the existence of robust chaos related to the parameter family.

Finally, when the maps $f$ and $g$ commute, the paradox can be impossible. For instance, it is proved in [Cánovas & Linero, 2005] that if $f$ is transitive (having a dense orbit) and piecewise monotone, then for any non constant continuous map $g$ commuting with $f$ it is held that $f \circ g$ is also transitive. Hence, the Parrondo's paradox is not possible. On the other hand, if both maps $f$ and $g$ commute and are piecewise monotone, then

$$\begin{cases} h(f \circ g) = \lim_{n\to\infty} \frac{1}{n} \log c((f \circ g)^n) \\ \quad = \lim_{n\to\infty} \frac{1}{n} \log c((f^n \circ g^n)) \\ \quad \leq \lim_{n\to\infty} \frac{1}{n} \log(c(f^n)c(g^n)) \\ = \lim_{n\to\infty} \frac{1}{n} \log c(f^n) + \lim_{n\to\infty} \frac{1}{n} \log c(g^n) = h(f) + h(g) \end{cases} \tag{9.26}$$

Hence, if $h(f) = h(g) = 0$ and both maps $f$ and $g$ are simple, then the composition $f \circ g$ is simple, and the Parrondo's paradox is not possible.

Our last open problem in this section is related to avoid piecewise monotonicity condition as follows:

**Problem 9.3.** *Let $f$ be continuous and transitive. Is it true that for any non constant continuous map $g$ commuting with $f$, the map $f \circ g$ is transitive? Can transitivity be replaced by another dynamic property in such a way an analogous result will be true?*

In the case of transitive piecewise monotone maps, the key for proving that $f \circ g$ is also transitive when $g \in C(I, I)$ commutes with $f$ is the following: for any compact subinterval $J \subset I$ there is a natural number $n(J)$ such that $f^{n(J)}(J) = I$. If $f$ is not piecewise monotone, we have that $\lim_{n\to\infty} f^n(J) = I$ and hence we think that we can avoid the assumption on the finite monotonicity pieces of $f$. The second part of the problem needs an exhaustive study of different dynamical properties, which is still undone.

### 9.3.2. *Dynamics Shared by Commuting Functions*

Commuting continuous maps defined on a compact metric space must have some dynamics in common. It was showed in [Hu, 1993] that, if $f$ and $g$ commute, then they share an invariant measure. Recall that an invariant measure of a continuous map $f$ is a probabilistic measure $\mu$ defined on the Borel sets of X, $\beta(X)$, such that for any $A \in \beta(X)$ we have that $\mu(A) = \mu(f^{-1}(A))$. However, the nature of such invariant measure is unclear.

Now, we focuss our attention on the set of continuous maps on the interval, and we wonder about the shared dynamics by two commuting continuous maps $f, g : [0,1] \to [0,1]$. The question that if $\mathrm{Fix}(f) \cap \mathrm{Fix}(g)$

was nonempty was open for a long time [Isbell, 1957], and finally was solved in [Boyce, 1969] and [Huneke, 1969], by finding two continuous commuting interval maps which do not share any fixed point.

In the general setting of compact metric spaces, it is well-known that $\text{Rec}(f) \cap \text{Rec}(g) \neq \emptyset$ (see [Furstenberg, 1981]). This result can be improved for one dimensional dynamics. Following the Sharkovsky's order of natural numbers, let $\mathcal{T}_1 = \{f \in C(I,I) : \text{Per}(f) \text{ is closed}\}$, $\mathcal{T}_2 = \{f \in C(I,I) : f$ has periodic points of period $2^n$, $n \geq 0\}$ and $\mathcal{T}_3 = \{f \in C(I,I) : f$ has a periodic point which is not a power of two$\}$. The following result stated in [Cánovas & Linero, 2005] shows that two commuting interval maps have to share some dynamics.

**Theorem 9.4.** *Assume $f, g \in C(I,I)$ commute. Then*

*(a) If $f \in \mathcal{T}_1$, then* $\text{Fix}(f) \cap \text{P}(g) \neq \emptyset$.
*(b) If $f \in \mathcal{T}_2$, then* $\text{Fix}(f) \cap \text{AP}(g) \neq \emptyset$.
*(c) If $f \in \mathcal{T}_3$, then* $\text{Fix}(f) \cap \text{UR}(g) \neq \emptyset$.

The key for proving Theorem 9.4 is that the set of fixed points of $f$ is invariant by $g$. The structure of different subsets of $f$ which are generated by its recurrence properties will give the role that fixed points of $f$ can play in the dynamics of $g$. For instance, if $f$ is piecewise monotone, then $\text{Fix}(f)$ is a finite union of (probably degenerate) compact subintervals of $I$, and hence $g$ has to share a fixed point with $f$, i.e., $\text{Fix}(F) \cap \text{Fix}(g) \neq \emptyset$. Since $\text{Fix}(f)$ is compact and invariant by $g$, it contains a minimal set, and since points in minimal sets are uniformly recurrent, we receive that at least $\text{Fix}(f) \cap \text{UR}(g) \neq \emptyset$. Our open problem concerning periodic points is the following one:

**Problem 9.4.** *Assume that $f$ and $g$ are commuting continuous interval maps. Since the counterexamples constructed in [Boyce, 1969] and [Huneke, 1969] the maps share periodic points, it is natural to raise the following question (see [Alikhani-Koopaei, 1998; Steele, 1998]): is it true that* $\text{Per}(f) \cap \text{Per}(g) \neq \emptyset$ *for commuting continuous interval maps? Note that, in view of Theorem 9.4, if $f$ and $g$ do not share a periodic orbit, the dynamical behaviors of both maps must be complicated. In particular, they must have infinitely many periodic points with an infinite $\omega$-limit set. Otherwise, we can apply Theorem 9.4 (a) to find shared periodic points.*

The next problem was suggested us by Professor Lubomir Snoha, from Matej Bel University.

**Problem 9.5.** *As we mentioned above, Fustenberg's result (see [Furstenberg, 1981]) states that two continuous commuting maps $f$ and $g$, which are defined on a general compact metric space, satisfy that* $\mathrm{Rec}(f) \cap \mathrm{Rec}(g) \neq \emptyset$. *Can it be improved by proving that* $\mathrm{UR}(f) \cap \mathrm{UR}(g) \neq \emptyset$? *Since the $\omega$-limit sets generated by uniformly recurrent points are minimal, the above question can be established in terms of minimal sets as follows: do commuting continuous maps share some minimal set?*

We have not reasonable conjectures concerning the above two questions, even when it is clear that commuting continuous maps have some dynamics in common because they share an invariant measure. However, the smallest set with full measure is the set of recurrent points.

### 9.3.3. *Computing Problems for Large Periods T*

As we pointed out in the introduction, sometimes the dynamical properties of a periodic sequence $f_{1,\infty}$ can be studied from dynamical systems properties of $f_T \circ ... \circ f_1$, $f_1 \circ f_T \circ ... \circ f_2, ...$ and $f_{T-1} \circ ... f_1 \circ f_T$. However, even in that case, there are some computing problems when the period $T$ is big enough. For instance, we introduce the following problem that was previously stated in [Elaydi, 2004].

**Problem 9.6.** *It is well-known that the logistic family* $f_\mu(x) = \mu x(1-x)$, $x \in [0,1]$ *and* $\mu \in [1,4]$, *has a fixed point which is an attractor for all the orbits from points* $x \in (0,1)$ *in the case that* $1 \leq \mu \leq 3$. *It is an open question to check the conditions on the parameters* $\mu_i$, $i = 1, ..., T$, *such that the periodic sequence* $(f_{\mu_1}, ..., f_{\mu_T}, ...)$ *has a periodic orbit of period $T$ which also is an attractor for all the trajectories in* $(0,1)$. *Of course, this periodic orbit* $(x_1, x_2, ..., x_T, ...)$ *has to satisfy*

$$|f'_{\mu_1}(x_1) f'_{\mu_2}(x_2) ... f'_{\mu_T}(x_T)| < 1$$

*but the family of parameters which makes possible the above equality is very difficult to characterize.*

The technical problems for solving the above problem starts when we try to find the fixed orbit of $f_{\mu_T} \circ ... \circ f_1$, because we have to solve the polynomial equation $f_{\mu_T} \circ ... \circ f_1(x) = x$. In general, it is well-known that there is no a closed formula for solving the above equation if the polynomial degree is high enough. So, alternative methods needs to be developed to get an useful approach to this problem.

### 9.3.4. *Commutativity Problems*

The dynamical properties of maps $f \circ g$ and $g \circ f$ need not be the same. For instance, one composition can be transitive (there exists a dense orbit) while the other composition is not surjective (recall that transitivity implies surjectivity), as the following example shows. Consider the maps

$$f(x) = \frac{1}{3}x + \frac{1}{3} \tag{9.27}$$

and

$$g(x) = \begin{cases} 0 & \text{if } x \in [0, \frac{1}{3}] \\ 6x - 2 & \text{if } x \in [\frac{1}{3}, \frac{1}{2}] \\ -6x + 4 & \text{if } x \in [\frac{1}{2}, \frac{2}{3}] \\ 0 & \text{if } x \in [\frac{2}{3}, 1] \end{cases} \tag{9.28}$$

It is well known that $(g \circ f)(x) = 1 - |1 - 2x|$ is transitive (see e.g. [Block & Coppel, 1992]), while

$$(f \circ g)(x) = \begin{cases} \frac{1}{3} & \text{if } x \in [0, \frac{1}{3}] \\ 2x - \frac{1}{3} & \text{if } x \in [\frac{1}{3}, \frac{1}{2}] \\ -2x + \frac{5}{3} & \text{if } x \in [\frac{1}{2}, \frac{2}{3}] \\ \frac{1}{3} & \text{if } x \in [\frac{2}{3}, 1] \end{cases} \tag{9.29}$$

and $f \circ g$ is not transitive since $(f \circ g)([0, 1]) = [1/3, 2/3]$.

In addition, the topological sequence entropies of $f \circ g$ and $g \circ f$ do not agree in general (see [Balibrea, 1999]). Topological sequence entropy is an extension of topological entropy defined from an strictly increasing sequence of positive integers $A = (a_n)_{n=0}^{\infty}$ as follows. Given a continuous map $f$, we define the sequence of maps $f_A = (f^{a_0}, f^{a_1 - a_0}, f^{a_2 - a_1}, ...)$. Then, the topological sequence entropy of $f$ relative to $A$ is the number $h_A(f) = h(f_A)$.

However, many dynamical properties are the same for both composition maps (see e.g. [Kolyada & Snoha, 1996, Cánovas *et al.*, 2006]). More precisely, a dynamical property $P$ is surjectivity independent if it satisfies the conditions:

(1) For any map $f \in C(I, I)$ it is held that $P \in \mathcal{D}(f)$ if and only if $P \in \mathcal{D}(f|_Y)$ where $Y = \cap_{n \geq 0} f^n(I)$.
(2) Given two arbitrary subintervals $J, K \subseteq I$, for any $f \in C(K, K)$, any $g \in C(J, J)$ and any surjective map $\varphi \in C(K, J)$ such that $\varphi \circ f = g \circ \varphi$, then either $P \in \mathcal{D}(g)$ implies $P \in \mathcal{D}(f)$ or $P \in \mathcal{D}(f)$ implies $P \in \mathcal{D}(g)$.

The list of dynamical properties which hold the above conditions includes, among others, positivity of topological entropy and chaoticity in the sense of Li-Yorke. Notice that, for instance, transitivity is not a surjectivity independent dynamical property. Then, the following result can be proved.

**Theorem 9.5.** *Let $f, g \in C(I, I)$ and assume $P$ is a dynamical property which is surjectivity independent. Then $P \in \mathcal{D}(f \circ g)$ if and only if $P \in \mathcal{D}(g \circ f)$.*

Now, we consider three continuous maps $f_i : I \to I$, $i = 1, 2, 3$. Then, Theorem 9.5 can be extended in such that way that if $P$ is a dynamical property which is surjectivity independent, then $P \in \mathcal{D}(f_1 \circ f_2 \circ f_3)$ if and only if $P \in \mathcal{D}(f_2 \circ f_3 \circ f_1)$ if and only if $P \in \mathcal{D}(f_3 \circ f_1 \circ f_2)$. However, we can see that $h(f_1 \circ f_2 \circ f_3) \neq h(f_2 \circ f_1 \circ f_3)$ (see [Kolyada & Snoha, 1996]). Just consider $I_1 = [0, 1/2]$ and $I_2 = [1/2, 1]$ and recall that $\varphi_{I_i} : I_i \to [0, 1]$ are increasing linear maps for $i = 1, 2$. Let $f(x) = 1 - |2x - 1|$ be the tent map and $g(x) = |2x - 1|$. It is well-known that $h(g) = h(f) = \log 2$.[a] Define the maps

$$f_1(x) = \begin{cases} x \text{ if } x \in I_1, \\ \frac{1}{2} \text{ if } x \in I_2, \end{cases} \tag{9.30}$$

$$f_2(x) = \begin{cases} \frac{1}{2} \quad \text{ if } x \in I_1, \\ 1 - x \text{ if } x \in I_2, \end{cases} \tag{9.31}$$

and

$$f_3(x) = \begin{cases} \varphi_{I_1} \circ f \circ \varphi_{I_1}^{-1}(x) + \frac{1}{2} \text{ if } x \in I_1, \\ 1 - x \qquad\qquad\qquad \text{if } x \in I_2. \end{cases} \tag{9.32}$$

Then

$$(f_1 \circ f_2 \circ f_3)(x) = \begin{cases} \varphi_{J_1} \circ g \circ \varphi_{J_1}^{-1}(x), & x \in I_1, \\ 1 - x, & x \in I_2, \end{cases} \tag{9.33}$$

and so $h(f_1 \circ f_2 \circ f_3) = \log 2$, while $f_2 \circ f_1 \circ f_3$ is the constant map equal to $\frac{1}{2}$, which has zero topological entropy.

Our last problem involves three continuous maps, is as follows:

**Problem 9.7.** *Find conditions on the maps such that a dynamical property holds for $f_1 \circ f_2 \circ f_3$ and $f_1 \circ f_3 \circ f_2$.*

[a]Note that both maps are piecewise monotone and $c(f^n) = c(g^n) = 2^n$.

Probably, in general the answer for this question will be negative, that is, for any dynamical property we expect the existence of three maps such that $f_1 \circ f_2 \circ f_3$ has such property and $f_1 \circ f_3 \circ f_2$ does not have it. But, notice that if $f_1$ is the identity, then $f_2 \circ f_3$ and $f_3 \circ f_1$ share a lot of dynamical properties. The question is finding another non trivial conditions on the maps in order both composition maps have the same dynamical properties.

If it is not allowed to the maps be defined piece by piece on subintervals, as for instance in analytical models, the above question can be reformulated as follows for the logistic family.

**Problem 9.8.** *Let $f_i(x) = a_i x(1 - x)$ be such that $a_i \in [1, 4]$. Is it true that a dynamical property holds for $f_1 \circ f_2 \circ f_3$ and $f_1 \circ f_3 \circ f_2$. Do their topological entropies agree?*

Here we have an additional problem because there is no way of computing topological entropy, and so, some numerical methods for computing topological entropy should be developed in advance.

### 9.3.5. *Generalization to Continuous Triangular Maps on the Square*

Continuous triangular maps on the square $T : [0, 1]^2 \to [0, 1]^2$ have the form

$$T(x, y) = (f(x), g(x, y)) \tag{9.34}$$

for all $(x, y) \in [0, 1]^2$. Note that the first coordinate is a continuous interval map, and this fact makes possible that Sharkovsky's theorem remains valid for this family of maps (see [Kloeden, 1979]). However, several authors have recently showed that the dynamics of triangular is richer than interval map dynamics (see e.g. [Forti, 1995, Kolyada, 1992; Snoha & Spitalsky, 2006] or [Alsedá *et al.*, 1993]). For instance, if $f \in C(I, I)$ holds that $\mathcal{S}(2^n)$, $n \in \mathbb{N} \cup \{\infty\}$, then its topological entropy is zero. However, there are triangular maps which do not hold this property. We can translate several problems to this family of maps as follows.

**Problem 9.9.** *Let $F$ be continuous and transitive triangular map. Is it true that for any non constant continuous map $G$ commuting with $F$, the map $F \circ G$ is transitive? Can transitivity be replaced by another dynamic property in such a way an analogous result is true?*

**Problem 9.10.** *Assume $F$ and $G$ are commuting continuous triangular maps. Is it true that* $\text{Per}(f) \cap \text{Per}(g) \neq \emptyset$*? Note that, since these maps have some common invariant measure, if both maps do not have infinite $\omega$-limit sets, they have to share some periodic orbit which is the support of the common invariant measure. So, if the property is false, the maps need be dynamically complicated.*

**Problem 9.11.** *Find for continuous triangular maps analogous results to Theorems 9.2 and 9.4. Note that triangular maps have periodic structure similar to Sharkovsky's interval result. So, this question makes sense.*

Triangular maps have a one dimensional first coordinate maps, and hence, the problems are expected with the second coordinate. Notice that any $x \in I$ generates a non-autonomous discrete system $f_{1,\infty} = (g_x, g_{f(x)}, g_{f^2(x)}, \ldots)$, where $g_x(y) = g(x, y)$ for all $(x, y) \in I$. Working with such non-autonomous discrete systems is extremely complicated and so, we do not have any reasonable conjecture.

## References


1. Adler, R. L., Konheim.A. G. & McAndrew, M. H. [1965] "Topological entropy," *Trans. Amer. Math. Soc* **114**, 309–319.
2. Alikhani-Koopaei, A. [1998] "On common fixed points and periodic points of commuting functions," *Internat. J. Math. & Math. Sci* **21**, 269–276.
3. Almeida, J., Peralta–Salas, D. & Romera, M. [2005] "Can two chaotic systems give rise to order?," *Phys. D* **200**, 124–132.
4. Alsedá, L., Kolyada. S. F. & Snoha, L. [1993] "On topological entropy of triangular maps of the square," *Bull. Austral. Math. Soc* **48**, 55–67.
5. Alsedá, L., Llibre.J. & Misiurewicz, M. [1993] *Combinatorial dynamics and entropy in dimension one*, World Scientific, Singapore.
6. AlSharawi, Z., Angelos, J., Elaydi, S. & Rakesh, L. [2006] "An extension of Sharkovsky's theorem to periodic difference equations," *J. Math. Anal. Appl* **316**, 128–141.
7. Andrecut, M. & Ali, M. K. [2001a] "Robust chaos in a smooth system," *Modern. Phys. Lett. B* **15**, 177–189.
8. Andrecut, M. & Ali, M. K. [2001b] "On the ocurrence of robust chaos in a smooth system," *Modern. Phys. Lett. B* **15**, 391–395.
9. Balibrea, F., Cánovas, J. S. & Jiménez López, V. [1999] "Commutativity and non-commutativity of the topological sequence entropy," *Ann. Inst. Fourier (Grenoble)* **49**, 1693–1709.
10. Blanchard, F., Glasner, E., Kolyada, S. & Maass, A. [2002] "On Li–Yorke pairs," *J. Reine Angew. Math* **547**, 51–68.
11. Block, L. & Coppel, W. A. [1992] *Dynamics in one dimension*, Lecture Notes in Mathematics **1513**, Springer-Verlag, Berlin.

12. Boyarsky, A., Gora, P. & Islam, M. S. [2005] "Randomly chosen chaotic maps can give rise to nearly ordered behavior," *Phys. D* **210**, 284–294.
13. Boyce, W. M. [1969] "Commuting functions with no common fixed point," *Trans. Amer. Math. Soc* **137**, 77–92.
14. Cánovas, J. S. & Linero, A. [2005] "On the dynamics of composition of commuting interval maps," *J. Math. Anal. Appl* **305**, 847–858.
15. Cánovas, J. S. & Linero, A. [2006] "Periodic structure of alternating continuous interval maps," *J. Difference Equ. Appl* **12**, 847–858.
16. Cánovas, J. S., Linero, A. & Peralta-Salas, D. [2006] "Dynamic Parrondo's paradox," *Phys. D* **218**, 177–184.
17. Cushing, J. M. & Henson, S. M. [2002] "A periodically forced Beverton–Holt equation," *J. Difference Equ. Appl* **8**, 1119–1120.
18. Dana, R. A. & Montrucchio, L. [1986] "Dynamic complexity in duopoly games," *J. Economic Theory* **44**, 40–56.
19. Devaney, R. L. [1989] *An introduction to chaotic dynamical systems*, Addison-Wesley, Redwood City.
20. Du, B. S.[2004] "A simple proof of Sharkovsky's theorem," *Amer. Math. Monthly* **111**, 595–599.
21. Elaydi, S. N. [2004] "Nonautonomous difference equations: open problems and conjectures," *Fields. Inst. Commun* **42**, 423–428.
22. Elaydi, S. N. & Sacker, R. J. [2005] "Nonautonomous Beverton–Holt equations and the Cushing-Henson conjectures," *J. Difference Equ. Appl* **11**, 337–346.
23. Forti, G. L., Paganoni, L. & Smítal, J. [1995] "Strange traingular maps of the square," *Bull. Austral. Math. Soc* **51**, 395–415.
24. Furstenberg, H. [1981] *Recurrence in ergodic theory and combinatorial number theory*, Princeton University Press, Pricenton, New Jersey.
25. Harmer, G. P. & Abbott, D. [1999] "Losing strategies can win by Parrondo's paradox," *Nature* **402**, 864.
26. Harmer, G. P. & Abbott, D. [1999] "Parrondo's paradox," *Stat. Sci* **14**, 206–213.
27. Hu, H. [1993] "Some ergodic properties of commuting diffeomorphisms," *Ergod. Th. and Dynam. Sys* **13**, 73–100.
28. Huneke, J. P. [1969] "On common fixed points of commuting functions on an interval," *Trans. Amer. Math. Soc* **139**, 371–381.
29. Isbell, J. R. [1957] "Commuting mappings of trees," *Bull. Amer. Math. Soc* **63**, 419.
30. Kloeden, P. E. [1979] "On Sharkovsky's cycle coexistence ordering," *Bull. Austral. Math. Soc* **20**, 171–177.
31. Kopel, M. [1996] "Simple and complex adjustment dynamics in Cournot duopoly games," *Chaos, Solitons & Fractals* **7**, 2031–2048.
32. Kolyada, S. F. & Snoha, L. [1996] "Topological entropy of nonautononous dynamical systems," *Random and Comp. Dynamics* **4**, 205–233.
33. Kopel, M. [1996] "Simple and complex adjustment dynamics in Cournot duopoly games," *Chaos, Solitons & Fractals* **7**, 2031–2048.
34. Misiurewicz, M. & Szlenk, W. [1980] "Entropy of piecewise monotone map-

pings," *Studia Math* **67**, 45–63.
35. Parrondo, J. M. R., Harmer, G. P. & Abbott, D. [2000] "New paradoxical games based on Brownian ratchets," *Phys. Rev. Lett* **85**, 5226–5229.
36. Puu, T. [1991] "Chaos in duopoly pricing," *Chaos, Solitons & Fractals* **1**, 573–581.
37. Puu, T. [2005] "Layout of a new industry: from oligopoly to competition," *Pure. Maths & Applications* **16**, 475–492.
38. Puu, T. & Norin, A. [2003] "Cournot duopoly when the competitors operate under capacity constraints," *Chaos, Solitons & Fractals* **18**, 577–592.
39. Sharkovsky, A. N. [1964] "Coexistence of cycles of a continuous transformation of a line into itself," *Ukrain. Mat. Zh* **16**, 61–71 (in Russian).
40. Smítal, J. [1986] "Chaotic functions with zero topological entropy," *Trans. Amer. Math. Soc* **297**, 269–282.
41. Snoha, L. & Spitalsky, V. [2006] "Recurrence equals uniform recurrence does not imply zero entropy for triangular maps of the square," *Discrete Contin. Dyn. Syst* **14**, 821–835.
42. Spurgin, R. & Tamarkin, M. [2005] "Switching investments can be a bad idea when Parrondo's paradox applies, *J. Behavioural Finance* **6**, 15–18.
43. Steele, T. H. [1998/9] "A note on periodic points and commuting functions," *Real Analysis Exchange* **24**, 781–790.
44. Wolf, D. M., Vazirani, V. V. & Arkin, A. P. [2005] "Diversity in times of adversity: probabilistic strategies in microbial survival games," *J. Theor. Biol.* **234**, 227–253.

# Chapter 10

# Code Structure for Pairs of Linear Maps with Some Open Problems

Pavel Troshin

*Department of Geometry, Faculty of Mechanics and Mathematics*
*Kazan State University*
*18 Kremlyovskaya str., Kazan, 420008, Russia*
*E-mail: Paul.Troshin@gmail.com*

Attractor of pair of complex linear maps together with the Mandelbrot set associated with this iterated function system were introduced by Barnsley and Harrington in 1985. Since that time many interesting properties of these sets have been found. We continue the investigation of this attractor by computing its code structure, the notion introduced by Barnsley in 2005.

**Keywords**: Pair of complex linear maps, code structure.

## Contents



## 10.1. Introduction

Iterated function systems (IFSes) are a particular case of multi–valued dynamical systems with contracting transformations, they take a special place among dynamical systems. IFSes represent the field of fractal geometry and thus combine both the simplicity of original construction and the richness of inner structure of their attractors which are, in some sense, self-similar sets. IFSes are of big practical and theoretical interest in coding and data reduction theory, computer graphics, modeling of natural objects, and in more abstract fields such as metric theory of numbers.

Particularly, we are interesting to IFSes given by pairs of complex linear maps. Such IFSes were first considered by Barnsley & Harrington [1985] together with the *Mandelbrot set $\mathcal{M}$ for pairs of linear maps.* Since that time these objects have been studied by Barnsley [1988], Bousch [1993], Indlekofer *et al.* [1993,1995], Solomyak [1998], Bandt [2002], Solomyak & Xu [2003], Solomyak [2005], Bandt & Hung [2008].

The current paper investigates the structure of attractor of pair of linear maps by computing its *code structure*, a new notion presented by Barnsley [2005, 2006]. Some open questions are listed in the end of this paper.

## 10.2. Iterated Function System

Hutchinson [1981] revived the idea of multi-valued functions and their invariant sets and proposed a simple approach to use this idea in clear and efficient way. Using his construction Barnsley & Demko [1985] and Barnsley [1988] developed the notion of *Iterated Function System* (IFS) and its applications, especially in building of fractal sets.

IFS $\{X; f_j\colon X \to X, j = 0, \ldots, m\}$ consists of a complete metric space $X$ and a finite family of contracting maps $f_j$ acting on it. Such dynamical system possesses its attractor — an invariant nonempty compact set $A = \bigcup_{j=0}^{m} f_j(A) \subset X$. One can also obtain the attractor $A$ by iterating maps $f_j$ applied to an arbitrary initial nonempty compact set (for instance single point $x \in X$):

$$A = \bigcup_{\sigma=\sigma_1\sigma_2\ldots\in\Omega} \lim_{n\to\infty} f_{\sigma_1} \circ f_{\sigma_2} \circ \ldots \circ f_{\sigma_n}(x)$$

where $\Omega = \{0, 1, \ldots, m\}^{\mathbb{N}}$ — is the set of sequences of symbols 0, 1, ..., $m$.

Most of the investigations in the field of IFSes deal with the case when all $f_j$ are similitudes on $\mathbb{R}^n$. In this case, the attractor consists of its scaled-down copies on every level of scale and thus usually it is a self-similar set with non-integer Hausdorff dimension.

Due to self-similarity of their attractors and simplicity of basic implementation rules (or axioms) IFSes are successfully used to *model many natural objects* such as boughs of trees, mountain landscapes, blood-vascular system, lungs, etc. In *computer graphics* IFSes help drawing fascinating fine pictures with multitude of details replicating themselves on each level of scale down to infinity (see [Barnsley, 1988]). To produce these objects (or images) with intricate structure one just needs to know the maps $f_j$ which often can be represented with a few bits of information. For instance plenty of desired attractors on the plane are given by IFSes consisting of affine maps, and each of these maps can be coded by 6 real numbers. Thus, it is enough to store just $6m$ numbers to produce fine pictures, using fast iterative algorithms. However there is an open question applicable to *coding and data reduction theory*: how to find efficiently affine maps which yield given picture as their attractor (see [Fisher, 1995]).

An example from *metric number theory* is the so-called $\beta$-expansions (introduced by Rényi [1957)) and problem of infinite Bernoulli convolutions connected to them (see Sec. 6.7 for more details). For $\beta \in (0,2]$ a sequence $\sigma = \sigma_0\sigma_1 \ldots \in \Omega = \{0,1\}^{\mathbb{N}}$ is called the $\beta$-expansion of $x \in [0,1]$ if

$$x = \sum_{j=1}^{\infty} \sigma_j \beta^{-j}$$

Estimation of all $\beta$-expansions of point $x$ as well as finding a unique top $\beta$-expansion (or *top address*, see Sec. 10.4) are connected to consideration of the IFS $\{[0,1]; \phi_0(x) = qx, \phi_1(x) = qx + 1 - q, \frac{1}{2} \le q = \frac{1}{\beta} < 1\}$.

Finally, we recall some facts about the Hausdorff dimension $\dim_H$ of attractor $A$ (see [Hutchinson, 1981] for more details). Fix $n \in \mathbb{N}$, $k \ge 0$ and let $\mathcal{H}^k$ be $k$-dimensional Hausdorff measure:

$$\mathcal{H}^k(E) = \sup_{\delta \ge 0} \inf \left\{ \sum_{i=1}^{\infty} (diam E_i)^k : \quad E \subset \bigcup_{i=1}^{\infty} E_i, \quad diam E_i \le \delta \right\}, \qquad E \subset R^n$$

A number $\dim_H \ge 0$ is called the Hausdorff dimension of $A$ if $\mathcal{H}^\alpha(A) = \infty$ for any $\alpha < \dim_H$ and $\mathcal{H}^\alpha(A) = 0$ for any $\alpha > \dim_H$.

Suppose $f_j \colon \mathbb{R}^n \to \mathbb{R}^n$ are similitudes with contraction ratios $0 < r_j < 1$, then the *similarity dimension* of $A$ is a number $D$ derived from the equation

$$\sum_{j=0}^{m} r_j^D = 1$$

and if $r_j = r$ for $j = 0, \ldots, m$, then

$$D = -\frac{\ln(m+1)}{\ln r}$$

It is rather simple to estimate $\dim_H(A)$, when $A$ is a self-similar set. This means that $f_j$ are similitudes with $0 < \mathcal{H}^{\dim_H}(A) < \infty$ and zero Hausdorff measure of each intersection of overlapping pieces: $\mathcal{H}^{\dim_H}(f_k(A) \cap f_l(A)) = 0$ $(l, k = 0, \ldots, m,\ l \ne k)$. In this case $\dim_H(A) = D$.

To ensure that pieces of $A$ do not overlap too much it is enough to check *Open Set Condition* (OSC). IFS satisfies OSC if there exists a non-empty open set $V \subset \mathbb{R}^n$ such that

(1) $\bigcup_{j=0}^{m} f_j(V) \subset V$,

(2) $f_k(V) \cap f_l(V) = \emptyset$ if $k \neq l$.

If OSC holds, then $A$ is self-similar set and $\dim_H(A) = D$. In [Bandt & Rao, 2007] there are also other useful equivalent statements of OSC (there exist weaker statements as well).

## 10.3. Attractor of Pair of Linear Maps

A pair of linear maps on $\mathbb{C}$ defines the iterated function system $\{\mathbb{C}; f_0, f_1\colon \mathbb{C} \to \mathbb{C}\}$,

$$\begin{cases} f_0(z) = qz \\ f_1(z) = qz + 1 \end{cases} \tag{10.1}$$

where $q \in \mathbb{C}$, $0 < |q| < 1$, is a fixed parameter. This IFS possesses its attractor - nonempty compact subset $A \subset \mathbb{C}$ such that $f_0(A) \cup f_1(A) = A$. This article concerns an approach to investigation of attractor $A$ by means of *code structure.*

These are the most general properties we need to know about attractor $A$:

- $A$ is a centrosymmetrical set [Bandt, 2002];
- $A$ is connected whenever $D(q) = f_0(A) \cap f_1(A) \neq \emptyset$ [Barnsley & Harrington, 1985];
- If $D(q) = \emptyset$ then $A$ is a Cantor set[a];
- There is a continuum of such values of $q$ that the set $D(q)$ consists of $2^n$ elements for any $n = 0, 1, \dots$ and continuum of such values of $q$ that $D(q)$ is a Cantor set with Hausdorff dimension $\beta$ for any $\beta \in (0, 0.2]$ [Bandt & Hung, 2008];
- Attractors of IFSes $\{\mathbb{C}; f_0(z) = qz, f_1(z) = qz+1\}$ and $\{\mathbb{C}; \widetilde{f}_0(z) = qz+a, \widetilde{f}_1(z) = qz + b, a, b \in \mathbb{C}, a \neq b\}$ are similar. This lets us consider only the former IFS;
- Fig. 10.1 shows some "typical" views of attractor $A$.

---

[a] By *Cantor set* (*with parameter* $r \in (0, 0.5)$) we denote a set, homeomorphous to the attractor of IFS $\{\mathbb{R}; S_0(x) = rx, S_1(x) = rx + 1\}$. This attractor is a perfect nonempty, compact metrizable and totally disconnected topological space. Under the Brouwer theorem (see [Kechris, 1995]) a Cantor set is characterized by these conditions up to homeomorphism.

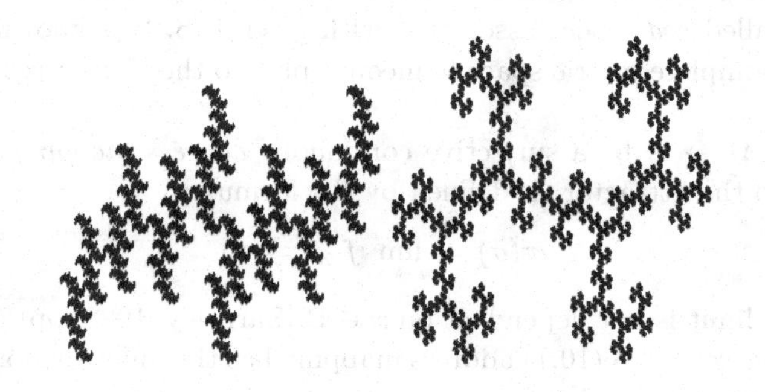

Fig. 10.1. Attractors of IFS (1): $A(q_1)$, $q_1 \approx 0.596 + 0.254i$ (on the left) and $A(q_2)$, $q_2 \approx 0.367 + 0.520i$ (on the right).

A *Mandelbrot set* $\mathcal{M}$ *for pairs of linear maps* is the set of values of parameter $q \in \mathbb{C}$ such that the attractor of IFS (10.1) is connected [Barnsley & Harrington, 1985]. The properties of this set together with many interesting discussions about it can be found in the cited literature, see also the last section of this paper.

## 10.4. Code Structure of Pair of Linear Maps

In order to investigate the structure of the attractor more thoroughly, we apply to this IFS a new notions of *top addresses* and *code structure* proposed in [Barnsley, 2005-2006]. Let $\Omega$ and $\Omega^*$ be correspondingly the sets of infinite and finite sequences of zeroes and ones (we will use bold Greek letters for them): $\sigma = \{\sigma_i\}_{i=0}^{k}$, $\sigma_i \in \{0,1\}$, $k = \infty$ or $k \in \mathbb{N}$, and let $|\sigma| = k+1$ be the length of sequence $\sigma \in \Omega^*$ (we also consider "empty" sequence with zero length). Let $\Omega_n^* = \{\sigma \in \Omega^* : |\sigma| = n\}$ and by $\sigma\omega = \sigma_0 \ldots \sigma_{|\sigma|-1}\omega_0\omega_1 \ldots$ we denote the concatenation of sequences $\omega \in \Omega$ and $\sigma \in \Omega^*$. Besides let $\sigma^n = \sigma \overset{n}{\ldots} \sigma$, $\sigma^\infty = \sigma\sigma\ldots$ and $f_{\sigma_0\ldots\sigma_n}(z) = f_{\sigma_0} \circ \ldots \circ f_{\sigma_n}(z)$.

Consider $\Omega$ endowed with metric $d_\Omega$ and linear ordering "$<$" as follows: let $\sigma, \omega \in \Omega$ and $k$ be the least number such that $\sigma_k \neq \omega_k$, then

$$\sigma < \omega \Leftrightarrow \sigma_k > \omega_k$$

$$d_\Omega(\sigma, \omega) = 2^{-k}$$

and $d_\Omega(\sigma,\omega) = 0$ whenever $\sigma = \omega$ (see details in [Barnsley, 2006]). The set $\Omega$ is called *code space* associated with given IFS. It is known that $\Omega$ is compact complete metric space homeomorphic to the Cantor set [Barnsley, 1988].

Let $\pi\colon \Omega \to A$ be a surjective continuous *address mapping* from code space onto the attractor $A$, defined by the formula:

$$\pi(\sigma) = \lim_{n\to\infty} f_{\sigma_0\ldots\sigma_n}(x) \tag{10.2}$$

where the limit is not dependent on $x \in \mathbb{C}$ [Barnsley, 1988, pp. 127].

In our case of IFS (10.1) address mapping has the following form [Barnsley & Harrington, 1985]:

$$\pi(\sigma) = \sum_{k=0}^{\infty} \sigma_k q^k \tag{10.3}$$

Indeed, one can derive Eq. (10.3) by passing to the limit in the following formula which is easily obtained by induction method: for any $n \in \mathbb{N}$, $\sigma \in \Omega_n^*$, $z \in \mathbb{C}$

$$f_{\sigma_0\ldots\sigma_{n-1}}(z) = q^n z + \sum_{k=0}^{n-1} \sigma_k q^k$$

Note that $\pi\colon \Omega \to A$ is injective if and only if $D(q) = f_0(A) \cap f_1(A) = \emptyset$ (such IFS is called *totally disconnected*). In this case, the attractor $A$ is a Cantor set [Barnsley, 1988, pp. 147]. We note that from the computability theory point of view it is undecidable to test if a given IFS is totally disconnected [Dube, 1994].

Following Barnsley [2005-2006] we note that the set[b] $\pi^{-1}(x) \equiv \{\sigma \in \Omega|\ \pi(\sigma) = x\}$ of addresses of $x \in A$ is closed and bounded by $0^\infty \in \Omega$ from above. Therefore, it contains the greatest element $\tau(x)$. Thus, we obtain the *function of top addresses* $\tau\colon A \to \Omega$ and $\tau(x) = \max\{\sigma \in \Omega|\ \pi(\sigma) = x\}$ is a *top address* of $x \in A$. Let $\Omega_\tau = \{\tau(x)|\ x \in A\} \subset \Omega$ be the set of all top addresses for a given IFS.

*A code structure* [Barnsley, 2006] is the following set of subsets of $\Omega$ (here the line over $\Omega_\tau$ denotes the closure):

$$\mathcal{C} = \{\pi^{-1}(x) \cap \overline{\Omega_\tau} : x \in A\}$$

The goal of this paper is to consider the code structure for the given IFS (10.1).

[b] For $x \in A$ we write $\pi^{-1}(x)$ instead of $\pi^{-1}(\{x\})$.

It is clear that if $D(q) = \emptyset$, then $\pi$ is injective and $\Omega_\tau = \Omega$, thus $\mathcal{C} = \{\pi^{-1}(x) : x \in A\} = \{\{\omega\} |\ \omega \in \Omega\}$.

Now we consider the case when $D(q)$ consists of $2^m$ elements, $m = 0, 1, \ldots$, or when $D(q)$ is a Cantor set. As it was shown in [Bandt & Hung, 2008] each of these cases takes place for appropriate continuum of values of $q$.

We call address $\omega \in \Omega$ of $x \in A$ *symmetrical to address* $\sigma \in \Omega$ of $y \in A$ if $\omega_k = 1 - \sigma_k$ for all $k = 0, 1, \ldots$. Thus, we denote by $\omega = \overline{\sigma}$ and $\omega_k = \overline{\sigma_k}$. Indeed, the points $x$ and $y$ are symmetrical relative to the point $z = \frac{1}{2(1-q)}$ because the symmetry mapping $s(z) = -z + (1-q)^{-1}$ turns $x$ into

$$s(x) = s(\pi(\omega)) = -\sum_{k=0}^{\infty} \omega_k q^k + (1-q)^{-1} = \sum_{k=0}^{\infty} (1 - \omega_k) q^k = \pi(\sigma) = y$$

Let $D(q) = f_0(A) \bigcap f_1(A)$. The structure of this intersection is studied in [Bandt & Hung, 2008], and we will use arguments similar to this paper.

If $D(q) \neq \emptyset$, then arbitrary point $x \in D(q)$ has at least two addresses: $\beta = 1\beta_1 \ldots$ and $\alpha = 0\alpha_1 \ldots$. These addresses fulfil the equation

$$\pi(\beta) - \pi(\alpha) = 1 + \sum_{k=1}^{\infty} (\beta_k - \alpha_k) q^k = 0 \tag{10.4}$$

By denoting $u_k = \beta_k - \alpha_k$, $u_k \in \{-1, 0, 1\}$ we obtain:

$$\sum_{k=0}^{\infty} u_k q^k = 0, u_0 = 1 \tag{10.5}$$

If we solve this equation with respect to $u_k$ (for given fixed $q$), we can restore all possible addresses of all points of $D(q)$. Indeed, if $u_k = 1$, then $\beta_k = 1$, $\alpha_k = 0$; if $u_k = -1$, then $\beta_k = 0$, $\alpha_k = 1$; if $u_k = 0$, then two cases are possible: $\beta_k = \alpha_k = 1$, $\beta_k = \alpha_k = 0$.

Consider $\mathbf{u} = 1w_1 0 w_2 0 w_3 0 \ldots$, where $w_i$ are words made of symbols $\{-1, 1\}$ (probably empty or infinite). Then using Eq. (10.4) we obtain

$$\begin{aligned} D(q) = \{\pi(\alpha) |\ \alpha = 0\widetilde{\omega}_1 v_1 \widetilde{\omega}_2 v_2 \ldots, \quad & v_i \in \{0, 1\}\} \\ = \{\pi(\beta) |\ \beta = 1\overline{\widetilde{w}}_1 v_1 \overline{\widetilde{w}}_2 v_2 \ldots, \quad & v_i \in \{0, 1\}\}, \end{aligned} \tag{10.6}$$

where $(\widetilde{w}_i)_j = 1$ if $(w_i)_j = -1$ and $(\widetilde{w}_i)_j = 0$ if $(w_i)_j = 1$, $j = 0, \ldots, |w_i| - 1$.

**Lemma 10.1.** *$D(q)$ is centrosymmetrical.*

***Proof.*** Since $A$ is symmetric, two different addresses of arbitrary point of $D(q)$ correspond to their two different correspondent symmetrical addresses of symmetrical point (probably coinciding with original point) thus also belonging to $D(q)$. □

Henceforward in this section we suppose that **Eq. (10.5) has unique solution u**. Examples of such values of $q$ can be found in Sec. 10.5 and two of them are depicted on Fig. 10.1 with $D(q) = \left\{\frac{1}{2(1-q)}\right\}$.

**Theorem 10.1.** *If there are $m \geq 0$ zeroes in $\mathbf{u}$, then $D(q)$ consists of $2^m$ different points $z_i$. Each point $z_i$ possesses exactly two addresses: $\alpha = 0\widetilde{w}_1 v_1 \widetilde{w}_2 v_2 \ldots$ and $\beta = 1\overline{\widetilde{w}}_1 v_1 \overline{\widetilde{w}}_2 v_2 \ldots$.*

***Proof.*** If $u_k = 0$ for finite set of indices $k = k_1, \ldots, k_m$, $m \geq 0$ only, then $D(q)$ consists of $2^m$ (probably coinciding) points $z_i$ with their pairs of addresses $(\beta^i, \alpha^i)$ (if $m = 0$, then $u_k \neq 0$ for any $k$). We will show that $z_i \neq z_j$, $i, j = 1, \ldots, 2^m$ $(i \neq j)$. Suppose the contrary, let $\beta^i \neq \beta^j$ both represent the same point, and these sequences differ from each other not more than in $m$ positions. Then

$$\pi(\beta^i) - \pi(\beta^j) = \sum_{k=1}^{\alpha(m)} \alpha_k q^k = 0$$

where $\alpha_k \in \{0, \pm 1\}$, and $\alpha(m)$ is the maximal position index of differing symbols in $\beta^i$ and $\beta^j$. Let $\alpha_{k_0}$ be the first non-zero coefficient (if all $\alpha_k = 0$, then $\beta^i = \beta^j$). We divide the equality by $\alpha_{k_0} q^{k_0}$ and thus obtain that the Eq. (10.5) has a solution $\mathbf{u}$ with infinite number of zeroes, which contradicts the assumption of the theorem. □

The following theorem and its proof are due to Bandt & Hung [2008], however we place it here for the sake of completeness.

**Theorem 10.2.** *If $u_k = 0$ for infinite number of indices $k = k_1 < k_2 < \ldots$ and $k_{n+1} - k_n > 2$ for $n$ big enough, then $D(q)$ also consists of infinite amount of points and all possible addresses $\alpha = 0\alpha_1\alpha_2 \ldots$ corresponding to $\mathbf{u}$ give us different points $z \in D(q)$. Every point of $D(q)$ has exactly two addresses.*

***Proof.*** Let us show that $\alpha \neq \alpha'$ implies $\pi(\alpha) \neq \pi(\alpha')$. Let $\widetilde{w} = 0\widetilde{w}_1 v_1 \ldots \widetilde{w}_n$, $\alpha = \widetilde{w} 1 \widetilde{w}_{n+1} v_{n+1} \ldots$ and $\alpha' = \widetilde{w} 0 \widetilde{w}_{n+1} v'_{n+1} \ldots$. Suppose that $\pi(\alpha) = \pi(\alpha')$, then we obtain:

$$\pi(\alpha) - \pi(\alpha')$$
$$= q^{k_n} + \sum_{p=n+1}^{\infty} (v_p - v'_p) q^{k_p} = q^{k_n} \left(1 + \sum_{p=n+1}^{\infty} (v_p - v'_p) q^{k_p - k_n}\right) = 0$$

From the uniqueness of $\mathbf{u}$ it follows that $u_k \neq 0$ is possible only when $k = k_p - k_n$, $p = n+1, \ldots$. Thus, the difference of sequence numbers of consecutive non-zero elements in $\mathbf{u}$ is greater than 2 when $k$ is big enough. Then the difference of sequence numbers of consecutive zeroes in $\mathbf{u}$ is not greater than 2. It contradicts the assumption of the theorem. □

Note that in this case $D(q)$ is a Cantor set [Bandt & Hung, 2008], and the authors state that the proof can be made without the assumption $k_{n+1} - k_n > 2$. Thus, we conclude that every point $z = \pi(0\widetilde{\omega}_1 v_1 \widetilde{\omega}_2 v_2 \ldots) \in D(q)$ can be denoted by $z_v = z_{v_1 v_2 \ldots}$ ($v \in \Omega$ or $v \in \Omega^*$). Recall that we use the notation $f_\sigma = f_{\sigma_0} \circ \ldots \circ f_{\sigma_n}$, $\sigma = \sigma_0 \ldots \sigma_n$.

**Lemma 10.2.** *For any $\sigma \in \Omega^*$ $(\sigma \neq \emptyset)$ and for any $z_1, z_2 \in D(q)$ $f_\sigma(z_1) \neq z_2$.*

***Proof.*** Let $f_\sigma(z_1) = z_2$ and let the point $z_1$ have two addresses $\beta$ and $\alpha$. Then the point $z_2$ has at least three different addresses: $\sigma\beta$, $\sigma\alpha$ and one beginning with $\overline{\sigma_1}$. This contradicts Theorems.10.1–10.2. □

**Corollary 10.1.** *For any $\sigma, \omega \in \Omega^*$ $(\sigma \neq \omega$ and $\sigma, \omega \neq \emptyset)$ and for any $z_1, z_2 \in D(q)$ , $f_\sigma(z_1) \neq f_\omega(z_2)$.*

***Proof.*** Let $\sigma = \sigma_0 \ldots \sigma_n$, $\omega = \omega_0 \ldots \omega_m$, $m \leq n$. Since $f_{\sigma_k \ldots \sigma_n}(z_1) \notin D(q)$ for any $k = \overline{0,n}$, $(k+1)^{th}$ symbol in the address of the point $f_\sigma(z_1)$ is uniquely defined and equal to $\sigma_k$ $(k = \overline{0,n})$. Thus, the addresses of the point $f_\sigma(z_1)$ begin with the word $\sigma$. We can say the same about addresses of point $f_\omega(z_2)$: they begin with the word $\omega$. Assuming that $f_\sigma(z_1) = f_\omega(z_2)$, we obtain $\omega_s = \sigma_s$, $s = \overline{0,m}$. Then, either $m = n$, which contradicts our settings, or $f_{\sigma_{m+1} \ldots \sigma_n}(z_1) = z_2$, which contradicts Lemma 10.2. □

Then, we can represent the attractor in the following way:

$$A = W \bigsqcup \left( \bigcup_{z \in D(q)} \bigcup_{\sigma \in \Omega^*} f_\sigma(z) \right) = W \bigsqcup V$$

where $W = \{x \in A |\ x \neq f_\sigma(z),\ \ \sigma \in \Omega^*,\ \ z \in D(q)\}$, $V = A \setminus W$, $\bigsqcup$ means disjoint union. Points from $W$ have unique addresses which are at the same time top addresses. Points from $V$ have exactly two addresses. Therefore, code space can be written as follows:

$$\Omega = \pi^{-1}(W) \bigsqcup \pi^{-1}(V) = \pi^{-1}(W) \bigsqcup \left( \bigcup_{\upsilon \in \widehat{\Omega}} \bigcup_{\sigma \in \Omega^*} \pi^{-1}(f_\sigma(z_\upsilon)) \right)$$

$$= \pi^{-1}(W) \bigsqcup \left( \bigcup_{\upsilon \in \widehat{\Omega}} \bigcup_{\sigma \in \Omega^*} \{\sigma 0 \widetilde{w}_1 v_1 \widetilde{w}_2 v_2 \ldots, \sigma 1 \overline{\widetilde{w}}_1 v_1 \overline{\widetilde{w}}_2 v_2 \ldots\} \right)$$

where $\widehat{\Omega} = \Omega^*_m$ in case when $D(q)$ is finite, and $\widehat{\Omega} = \Omega$ in the opposite case ($\widehat{\Omega} = \emptyset$, if $D(q) = \emptyset$).

Henceforward we use the notation $A^o$ for $Int(A)$ - interior of $A$, $\overline{A}$ for $Clos(A)$- closure of $A$, and $A^c$ for complement of $A$.

**Lemma 10.3.** $(\pi^{-1}(V))^o = \emptyset$.

***Proof.*** We denote $F = \pi^{-1}(V)$. Assuming the contrary and let $\xi \in F$ and $F \supset B(\xi, 2^{-k}) = \{\rho | \, d_\Omega(\xi, \rho) \leq 2^{-(k+1)}\}$ – an open ball of radius $2^{-k}$, where $k$ is an arbitrary large natural number.

If $D(q)$ is finite, then $F$ is countable, whereas $B(\xi, 2^{-k})$ is continuum, which is a contradiction.

If $D(q)$ is infinite we consider two addresses $\phi = x_1 \ldots x_{k-1} 0^\infty$ and $\varphi = x_1 \ldots x_{k-1}(10)^\infty$, $\phi, \varphi \in B(\xi, 2^{-k})$. Since $B(\xi, 2^{-k}) \subset F$, then using Eq. (10.6) we obtain $\phi = \sigma\alpha = \sigma 0 \widetilde{w}_1 v_1 \widetilde{w}_2 v_2 \ldots$. Which implies $\widetilde{w}_i = 0 \ldots 0$ ($0 \leq |\widetilde{w}_i| < \infty$ zeroes) for big enough $i \in \mathbb{N}$. Also note that $|\widetilde{w}_i| = k_{i+1} - k_i - 1 > 1$ when $i$ is big enough.

Similarly, $\varphi = \sigma'\alpha' = \sigma' 0 \widetilde{w}_1 v'_1 \widetilde{w}_2 v'_2 \ldots$, which implies that $|\widetilde{w}_j| \leq 1$, beginning with some big enough $j \in \mathbb{N}$. This leads to a contradiction with our assumptions. □

**Corollary 10.2.** *If* $D(q) \neq \emptyset$, *then* $\overline{V} = \overline{W} = A$, $V^o = W^o = \emptyset$.

***Proof.*** First we show that $V^o = \emptyset$. If $V^o \neq \emptyset$, then there exists a point $x \in V$ and its open neighbourhood $U(x) \subset V$. Then $\pi^{-1}(V) \supset \pi^{-1}(U(x))$ is an open set, since address mapping $\pi$ is continuous. This contradicts Lemma 10.3. Secondly, we show that $\overline{V} = A$. Let $z_1 \in D(q)$. An arbitrary point $W \ni x = \pi(\sigma) = \lim_{n\to\infty} f_{\sigma_0 \ldots \sigma_n}(z_1)$ can be approximated by points $x_n = f_{\sigma_0 \ldots \sigma_n}(z_1) \in V$ (see Eq. (10.2)). Since $A$ is closed, $\overline{V} = A$. Consider the topological subspace $A \subset \mathbb{C}$: $\overline{W} = \overline{V^c} = V^{oc} = \emptyset^c = A$, $W^o = W^{occ} = (\overline{W^c})^c = (\overline{V})^c = A^c = \emptyset$. □

**Lemma 10.4.** $\overline{\Omega_\tau} = \Omega$.

***Proof.*** Again denote $F = \pi^{-1}(V)$. Since $F^c \subset \Omega_\tau$, then $F^{oc} = \overline{F^c} \subset \overline{\Omega_\tau}$. However $F^o = \emptyset$, which implies $\Omega = F^{oc} \subset \overline{\Omega_\tau}$. □

Lemmas.10.3-10.4 imply that either when the solution $\mathbf{u}$ of Eq. (10.5) is unique, or when $D(q) = \emptyset$, the following theorem holds true.

**Theorem 10.3.** *Code structure for IFS (10.1) is as follows:*

$$\mathcal{C} = \{\pi^{-1}(x) : x \in W\} \bigsqcup \left( \bigcup_{z \in D(q)} \bigcup_{\sigma \in \Omega^*} \{\pi^{-1}(f_\sigma(z))\} \right)$$

$$= \{\pi^{-1}(x) : x \in W\} \bigsqcup \left( \bigcup_{v \in \widehat{\Omega}} \bigcup_{\sigma \in \Omega^*} \{\pi^{-1}(f_\sigma(z_v))\} \right)$$

$$= \{\pi^{-1}(x) : x \in W\} \bigsqcup \left( \bigcup_{v \in \widehat{\Omega}} \bigcup_{\sigma \in \Omega^*} \{\{\sigma 0 \widetilde{w}_1 v_1 \widetilde{w}_2 v_2 \ldots, \sigma 1 \overline{\widetilde{w}}_1 v_1 \overline{\widetilde{w}}_2 v_2 \ldots\}\} \right)$$

*where* $\widehat{\Omega} = \Omega_m^*$ *if* $D(q)$ *is finite, and* $\widehat{\Omega} = \Omega$ *in the opposite case* ($\widehat{\Omega} = \emptyset$ *if* $D(q) = \emptyset$).

## 10.5. Sufficient Conditions for Computing the Code Structure

We specially consider the case when $D(q)$ is a singleton (consists of one element). Since $A$ is symmetrical with respect to point $Z = \frac{1}{2(1-q)}$, then $D(q) = \{Z\}$.

**Lemma 10.5.** *If* $D(q) = Z$, *then* $u_k \neq 0$, $k = \overline{0, \infty}$, *where* $\boldsymbol{u} = u_0 u_1 \ldots$ *is a solution of Eq. (10.5).*

***Proof.*** Suppose that there exists $\tilde{k} > 0$ such that $u_{\tilde{k}} = 0$. Then the point $Z$ possesses two addresses $\alpha = 0\mu 1\nu$ and $\alpha' = 0\mu 0\nu$, $\mu \in \Omega^*_{\tilde{k}-1}$, $\nu \in \Omega$. Then $\pi(\alpha) - \pi(\alpha') = q^{\tilde{k}} = 0$, which contradicts $|q| > 0$. □

**Lemma 10.6.** *If* $D(q) = \left\{\frac{1}{2(1-q)}\right\}$, *then the Eq. (10.5) has unique solution* $\boldsymbol{u}$.

***Proof.*** Suppose that the Eq. (10.5) has two solutions $\mathbf{u}$ and $\mathbf{u}'$. Each solution corresponds to two addresses

$$\beta = \omega\beta_p\beta_{p+1}\ldots \text{ and } \beta' = \omega\beta'_p\beta'_{p+1}\cdots$$

of the point $Z$, where $\omega \in \Omega^*$, $|\omega| = p \geq 1$, $\beta_p = 1$, $\beta'_p = 0$. Thus

$$\pi(\beta) - \pi(\beta') = q^p + \sum_{n=p+1}^{\infty} (\beta_n - \beta'_n)q^n = q^p \left(1 + \sum_{n=p+1}^{\infty} (\beta_n - \beta'_n)q^{n-p}\right) = 0.$$

Since $\beta_k - \beta'_k \in \{0, \pm 1\}$, then $1(\beta_{p+1} - \beta'_{p+1})(\beta_{p+2} - \beta'_{p+2})\ldots$ is a solution of the Eq. (10.5). Thus, $\beta_k - \beta'_k \neq 0$ if $k = p+1, \ldots$. And this means that if $\sigma = \beta_{p+1}\beta_{p+2}\ldots$, then $\beta'_{p+1}\beta'_{p+2} = \overline{\sigma}$. Thus, $\beta = \omega\sigma$, $\beta' = \omega\overline{\sigma}$. Symmetrical address $\overline{\beta'} = \overline{\omega}\sigma$ is also an address of the point $Z$. Therefore, the following equation should hold true:

$$\pi(\beta) - \pi(\overline{\beta'}) = \sum_{k=0}^{p-1} (\omega_k - \overline{\omega_k})q^k = 0$$

Since $\omega_k - \overline{\omega_k} \neq 0$, we have obtained a solution of Eq. (10.5) which contains zeroes (infinite number) and this contradicts Lemma 10.5. $\square$

We note here that Solomyak [2005] proved more general result: $\mathbf{u}$ is unique in cases when $D(q)$ consists of 1 or 2 elements. This Lemma helps finding unique solutions of Eq. (10.5) and corresponding $q$ graphically, by drawing attractors with PC and writing down and solving equivalence between two addresses of point $Z$. However, usually it is not easy to verify the uniqueness of solution $\mathbf{u}$ (for given value of $q$).

Now using the argument from [Bandt & Hung, 2008] and [Solomyak, 2005] we show how to prove the uniqueness of $\mathbf{u}$ for some $q$. For this purpose we shall need the following a Lemma due to Bandt & Hung [2008] (we omit the proof of it).

**Lemma 10.7.** *Let $q$ and $\mathbf{u}$ satisfy Eq. (10.5). Let $U$ be simply connected neighbourhood of $q$. Then there exists $m \in \mathbb{N}$ such that for any other $\mathbf{u}' = 1u_1 \ldots u_m u'_{m+1} \ldots$ there exists $q' \in U$ satisfying Eq. (10.5).*

Consider the following set:

$$\mathfrak{W} = \{q \in \mathbb{C} |\arg q \in (20^o; 25^o), |q| \in (0.59; 0.66)\}$$

We note that $\mathfrak{W} \ni q_1 \approx 0.596 + 0.254i$ - the root of the equation $2q^3 - 2q + 1 = 0$ corresponding to $\mathbf{u}^1 = 1\text{-}1\text{-}11^\infty$. By Lemma 10.7 there exists such minimal

$m_0$ such that for any $\widetilde{\mathbf{u}} = 1\text{-}1\text{-}11^{m_0}u'_{m_0+3}\dots$ the solution of the Eq. (10.5) lies in $\mathfrak{W}$. Now consider the set

$$\mathfrak{D} = \left\{ g(z) = 1 - z - z^2 + \sum_{j=3}^{\infty} v_j z^j \;\middle|\; v_j = 1 \text{ for } j \le \max\{9, m_0+2\} \right.$$

$$\left. \text{and } v_j \in \{-1,0,1\} \text{ for other } j, \text{ in addition} \quad (v_i, v_j \ne 1) \Rightarrow |i-j| \ge 9 \right\}$$

**Lemma 10.8.** *For any $g \in \mathfrak{D}$ there exists $q \in \mathfrak{W}$ such that $g(q) = 0$ and Eq. (10.5) has unique solution $\mathbf{u}$. All such values of $q$ are different.*

***Proof.*** Using Lemma 10.7 it is enough to verify the uniqueness (this will imply that all $q$ are different). We use the idea of argument from [Bandt & Hung, 2008]. Fix $g \in \mathfrak{D}$, $g(q) = \sum\limits_{k=0}^{\infty} u_k q^k = 0$. Let $\mathbf{u}'$ be another solution of Eq. (10.5), and $p(q) = \sum\limits_{k=0}^{\infty} u'_k q^k = 0$. Then $g(q) - p(q) = 0$. Let $p(q) = \sum\limits_{k=0}^{k_0-1} u_k q^k + \sum\limits_{k=k_0}^{\infty} u'_k q^k$, $u_{k_0} \ne u'_{k_0}$, and $k_0 \ge 1$ be the least such number. Then

$$g(q) - p(q) = q^{k_0}\left(u_{k_0} - u'_{k_0} + \sum_{k=1}^{\infty}(u_{k+k_0+1} - u'_{k+k_0+1})q^k\right) = 0$$

$$\frac{u_{k_0} - u'_{k_0}}{q} + \sum_{k=0}^{\infty}(u_{k+k_0+1} - u'_{k+k_0+1})q^k = 0 \tag{10.7}$$

$$\frac{u_{k_0} - u'_{k_0}}{q^2} + \frac{u_{k_0+1} - u'_{k_0+1}}{q} + \sum_{k=0}^{\infty}(u_{k+k_0+2} - u'_{k+k_0+2})q^k = 0 \tag{10.8}$$

Let $\Im(a)$ be imaginary part of complex number $a \in \mathbb{C}$. We note that

$$\Im(q^{-1}) = -|q|^{-2}\Im(q) > -|q|^{-1}\sin 25^o > -\frac{0.43}{0.59} > -0.73$$

$$\Im(q^{-2}) = -|q|^{-4}\Im(q^2) < -|q|^{-2}\sin 40^o < -\frac{0.63}{0.59^2} < -1.8$$

$$\max_{k=\overline{1,6}} \Im(q^k) < \max\{0.66\sin 25^o, 0.66^2\sin 50^o, 0.66^3\sin 75^o, 0.66^4\sin 80^o,$$
$$0.66^5\sin 100^o, 0.66^6\sin 120^o\} < 0.334$$

and $\Im(q^k) > 0$ if $k = \overline{1,6}$. Also if $c_k \in \{-2, 0, 1, 2\}$, then

$$\left|\Im\left(\sum_{k=7}^{\infty} c_k q^k\right)\right| \le \left|\sum_{k=7}^{\infty} c_k q^k\right| \le 2\sum_{k=7}^{\infty} |q|^k = 2\frac{|q|^7}{1-|q|} < 0.33$$

Therefore, taking into account the structure of the set $\mathfrak{D}$, if $u_{k_0} - u'_{k_0} < 0$ ($u_{k_0} \in \{-1, 0\}$, $u_{k_0} - u'_{k_0} \in \{-1, -2\}$) 7 of the first coefficients of the series $\sum_{k=0}^{\infty} (u_{k+k_0+2} - u'_{k+k_0+2})q^k$ are nonnegative and

$$\Im\left(\frac{u_{k_0} - u'_{k_0}}{q^2} + \frac{u_{k_0+1} - u'_{k_0+1}}{q} + \sum_{k=0}^{\infty}(u_{k+k_0+2} - u'_{k+k_0+2})q^k\right)$$
$$> 1.8 - 2 \cdot 0.73 - 0.33 > 0.01$$

and if $u_{k_0} - u'_{k_0} > 0$ ($u_{k_0} \in \{0, 1\}$, $u_{k_0} - u'_{k_0} \in \{1, 2\}$), then among the first 6 coefficients of the series $\sum_{k=1}^{\infty} (u_{k+k_0+1} - u'_{k+k_0+1})q^k$ there are not more than one negative and

$$\Im\left(u_{k_0} - u'_{k_0} + \sum_{k=1}^{\infty}(u_{k+k_0+1} - u'_{k+k_0+1})q^k\right) > 1 - 2 \cdot 0.334 - 0.33 > 0.002$$

which contradicts Eqs. (10.7)–(10.8). □

**Corollary 10.3.** *Let $q \in \mathfrak{W}$, $g \in \mathfrak{D}$, $g(q) = 0$, and $\boldsymbol{u}$ be the solution of the Eq. (10.5) corresponding to $q$, $p(q) = \sum_{k=0}^{k_0-1} u_k q^k + \sum_{k=k_0}^{\infty} u'_k q^k$, $u_{k_0} \ne u'_{k_0}$, $u'_k \in \{-1, 0, 1\}$, and $k_0 \ge 1$ is the least such number. Then*

$$|p(q)| > 0.002|q|^{k_0}.$$

***Proof.*** Using the argument of the proof of Lemma 10.8, we obtain:

$$|p(q)| = |p(q) - g(q)| = |q^{k_0}|\left|\sum_{k=0}^{\infty}(u_{k+k_0} - u'_{k+k_0})q^k\right|$$
$$\ge |q|^{k_0}\left|\Im\left(\sum_{k=0}^{\infty}(u_{k+k_0} - u'_{k+k_0})q^k\right)\right|$$
$$> \min\{0.002, 0.01|q|^2\}|q|^{k_0} = 0.002|q|^{k_0}.$$ □

**Corollary 10.4.** *Let $q_1 \approx 0.596 + 0.254i$ be a root of equation $2q^3 - 2q + 1 = 0$. Then the Eq. (10.5) has unique solution $\boldsymbol{u}^1 = 1\text{-}1\text{-}11^{\infty}$.*

Note that if there are exactly $m \geq 0$ zeroes among the coefficients of $g \in \mathfrak{D}$, then the set $D(q)$ consists of $2^m$ elements, and if there is infinite number of zeroes, then $D(q)$ is not countable (more precisely, it is a Cantor set [Bandt & Hung, 2008]). In both cases any such $q$ corresponds to unique word $\mathbf{u}$. Thus the condition of uniqueness for the solution $\mathbf{u}$ of the Eq. (10.5) is not too much strict. Let $\dim_H(A)$ be the Hausdorff dimension of the set $A \subset \mathbb{C}$. Note that for any $0 < |q| < 1$ [Barnsley & Demko, 1985] $\dim_H(A) \leq -\frac{\ln 2}{\ln |q|}$. If the attractor $A$ is not connected ($f_0(A) \cap f_1(A) = \emptyset$), then

$$\dim_H(A) = -\frac{\ln 2}{\ln |q|} \tag{10.9}$$

If $A$ is connected ($q \in \mathcal{M}$ (the Mandelbrot set)), then it is not always easy to estimate $\dim_H(A)$. Bandt & Rao [Bandt & Rao, 2007] showed that if $f_0(A) \cap f_1(A)$ is finite but not empty, then "open set condition" (OSC) holds true. This condition is important for calculating Hausdorff dimension of fractals (in our case OSC means that there exists bounded open set $V \subset \mathbb{C}$ such that $f_0(V) \cup f_1(V) \subset V$, $f_0(V) \cap f_1(V) = \emptyset$ [Moran, 1946]). In [Bandt & Rao, 2007] there are also other useful equivalent statements of OSC. When OSC is fulfilled Hausdorff dimension coincides with "similarity dimension" (see [Hutchinson, 1981]) and can be calculated by Eq. (10.9). Some reasoning to use Eq. (10.9) in case when $D(q)$ is not countable can be found in the following result:

**Lemma 10.9.** *If $q \in \mathfrak{W}$, $g \in \mathfrak{D}$, $g(q) = 0$, then for IFS (10.1) OSC holds true.*

***Proof.*** We use the method invented in [Bandt & Hung, 2008]. Indeed, Bandt & Graf [1992] proved that OSC holds true if and only if the identity mapping is not a limit point of set of mappings $f_v^{-1} f_\omega$, where $v = v_1 \ldots v_m, \omega = \omega_1 \ldots \omega_m \in \Omega_m^*$, $m \in \mathbb{N}$, $v_1 \neq \omega_1$. It is also shown in that paper that our mappings are parallel translations:

$$f_v^{-1} f_\omega(z) = z + \sum_{j=1}^{m} \frac{1}{q^j}(\omega_j - v_j) = z + \frac{p(q)}{q^{m'}}$$

where $p(q) = \sum_{j=0}^{m'-1} b_j q^j$ is a polynomial of degree $b_j \in \{-1, 0, 1\}$, $b_0 \in \{-1, 1\}$, $1 \leq m' \leq m$. We need to show that the lengths of vectors of parallel translations are bounded from below by positive constant. Indeed, using the corollary from Lemma 10.8, we obtain

$$\left|\frac{p(q)}{q^{m'}}\right| = \frac{|g(q)-p(q)|}{|q|^{m'}} > 0.002|q|^{k_0-m'} \geq 0.002$$

where $g(q) = \sum_{k=0}^{\infty} u_k q^k = 0$, $k_0 \leq m'$ (since $b_j = 0$, and $u_j \neq 0$ if $j \geq m'$). □

Denote

$$\mathfrak{P} = \{q \in \mathfrak{M} |\ \exists g \in \mathfrak{D} : \ g(q) = 0\} \subset \mathbb{C}, \quad \mathfrak{P} \neq \emptyset$$

Recall that if the solution $\mathbf{u}$ of Eq. (10.5) is unique, then $A = W \bigsqcup V$, where $W$ consists of points of attractor having exactly one address, and $V$ consists of points of attractor having exactly two addresses. Besides $V = \bigcup_{z \in D(q)} \bigcup_{\sigma \in \Omega^*} f_\sigma(z)$. Now we summarize the results from Theorem 10.3 and Lemmas 10.8 and 10.9.

**Theorem 10.4.** *If Eq. (10.5) has unique solution* $\mathbf{u}$*, then the code structure for IFS (1) is as follows:*

$$\mathcal{C} = \{\pi^{-1}(x) : x \in W\} \bigsqcup \{\pi^{-1}(x) : x \in V\}$$

*If* $q \in \mathfrak{P}$*, then Eq. (10.5) has unique solution* $\mathbf{u}$ *and for the IFS (10.1) OSC holds true.*

Note that the structure of the set $\{\pi^{-1}(x) : x \in V\}$ is given more precisely in Theorem 10.3. Note also that $D(q) = \emptyset$ implies $\mathcal{C} = \{\{\omega\} |\ \omega \in \Omega\}$.

Another example of $q$, for which Eq. (10.5) has unique solution $\mathbf{u}$ and for IFS (10.1) OSC holds true is $q_2 \approx 0.367+0.520i$ (the root of $1-q+q^2+2q^3 = 0$, in this case $\mathbf{u}^2 = 1\text{-}1(111\text{-}1\text{-}1\text{-}1)^\infty$, see [Bandt & Hung, 2008]). Numbers $q_1$ and $q_2$ are special for they lie on the boundary of the Mandelbrot set $\mathcal{M}$ (see [Bandt, 2002]). Fig. 10.1 shows attractors $A(q_1)$ and $A(q_2)$ of IFS (10.1) when $q = q_1$ and $q = q_2$ correspondingly, $D(q) = \left\{\frac{1}{2(1-q)}\right\}$. Therefore $\dim_H(A(q_1)) = \log_{|q_1|} 0.5 \approx 1.596$, $\dim_H(A(q_2)) = \log_{|q_2|} 0.5 \approx 1.535$.

## 10.6. Conclusion and Open Questions

We would like to formulate some open questions concerning the attractor $A(q)$ and the Mandelbrot set associated with the given IFS (10.1). Some of these questions were asked in profound papers by Bandt [2002] and Solomyak [2005].

**Problem 10.1.** *We recall that the Mandelbrot set for pairs of linear maps is a set of values of parameter* $q \in \{q \in \mathbb{C} ||q| < 1\}$ *for which attractor*

*$A(q)$ of IFS (10.1) is connected. This set was established by Barnsley & Harrington [1985]. Their paper showed that $\mathcal{M}$ is symmetrical with respect to the real and imaginary axes and*

$$\{2^{-1/2} \leq |q| < 1\} \subset \mathcal{M} \subset \{1/2 \leq |q| < 1\}$$

*Bousch [1993] proved that $\mathcal{M}$ is connected and even locally connected. Bandt [2002] showed that $\mathcal{M}$ is not simply connected: there exist holes in it. Solomyak [2005] and Bandt & Hung [2008] found uncountable number of values of $q$ for which $D(q)$ is a singleton: $D(q) = \{1/(2-2q)\}$ (in this case (see [Bandt & Keller, 1991]) attractor $A(q)$ is a dendrite: it is connected, nowhere-dense compact set in $\mathbb{R}^2$ with connected complement).*

Thus, the open question proposed by Bandt [2002] is as follow:

- Is the set $\mathcal{M}\setminus\mathbb{R}$ regular-closed, that is the closure of its interior: $\mathcal{M}\setminus\mathbb{R} \subset clos(int(\mathcal{M}))$?

This problem stays open, though Solomyak & Xu [2003] proved this statement for some special neighbourhood of the imaginary axis.

**Problem 10.2.** *Since both intervals $[-1,-0.5]$ and $[0.5,1]$ lie in $\mathcal{M}$, it is natural to ask*

- *What is the smallest real $r > 0$ in the closure of $\mathcal{M} \setminus \mathbb{R}$?*

*Barnsley & Harrington [1985] proved $r \geq 0.53$. Solomyak [1998] showed that $r \geq 0.649$. With the help of computer figures Bandt [2002] states that $r \approx 0.6685 > 2/3$ and that calculations near the real axis become extremely time-consuming.*

**Problem 10.3.** *What is the point of minimum modulus in $\mathcal{M} \setminus \mathbb{R}$? Bandt [2002] and Solomyak [2005] notice it is possible that such point is situated near $q_2 \approx 0.367 + 0.520i$ (a zero of $1 - z + z^2 + 2z^3 = 0$) — a tip of a spiral sprout part in $\mathcal{M}$.*

**Problem 10.4.** *Barnsley [1988] and Bandt [2002] noticed a resemblance between local structure of $\mathcal{M}$ near its boundary and corresponding attractor $A(q)$, they asked if this connection can be made rigorous. Solomyak [2005] claims that there is much closer resemblance of $\mathcal{M}$ to the attractors $\widetilde{A}(q)$ of the IFS consisting of three contractions $\{qz-1, qz, qz+1\}$ and proves asymptotic similarity of $\mathcal{M}$ to $\widetilde{A}(q)$ for some 'landmark' points of $\mathcal{M}$. However the general question is still exists.*

- *How does $D(q)$ change with changing $q \in \mathcal{M}$?*

**Problem 10.5.** *We have showed that if $D(q)$ is a singleton, then there is a unique solution* $\mathbf{u}$ *of Eq. (10.5). There are some known values of $q \in \partial\mathcal{M}$ being in some sense 'landmark' points of $\mathcal{M}$: $q_1 \approx 0.596 + 0.254i$, $q_2 \approx 0.367+0.520i$, $q_3 \approx 0.290+0.586i$, $q_4 \approx 0.636+0.107i$ (see [Solomyak, 2005]). The corresponding sets $D(q_i)$ are singletons. For $q_5 \approx 0.574+0.369i$ and $q_6 \approx 0.622 + 0.188i$ the number of elements in $D(q)$ is 2, for $q_7 \approx 0.644 + 0.141i$ it is 4. Thus it is interesting to know*

- *Is it true that $D(q)$ is finite for "most points" $q \in \partial\mathcal{M}$?*
- *What is the connection between $q \in \partial\mathcal{M}$ and uniqueness of solution $\boldsymbol{u}$?*

**Problem 10.6.** *In the present paper we analyzed the case when the solution* $\mathbf{u}$ *of Eq.(10.5) is unique. The more complicated case is left to consider, in it every point of $D(q)$ can possess $2^k$ addresses, $k = 0, 1, \ldots$ and $k$ depends on number of solutions* $\mathbf{u}$ *and on numbers of zeroes in them. Despite this complexity it is still possible that $\overline{\Omega_\tau} = \Omega$. Thus we left the question*

- *Describe code structure in the case when the solution $\boldsymbol{u}$ is not unique.*

**Problem 10.7.** *Let us consider particular case, when $q \in (0, 1) \subset \mathbb{R}$. For every such $q$ consider the image of the Bernoulli product probability measure on $\Omega$ with $P(0) = P(1) = \frac{1}{2}$ under the projection $\pi$. The resulting measure $\nu_q$ is called Bernoulli convolutions. In more details, if $\mu = (\frac{1}{2}, \frac{1}{2})^{\mathbb{N}}$ is the Bernoulli measure on $\Omega = \{0, 1\}^{\mathbb{N}}$, then*

$$\nu_q = \mu \circ \pi^{-1}, \quad \text{where} \quad \pi(\sigma) = \sum_{k=0}^{\infty} \sigma_k q^k, \quad \sigma \in \Omega$$

*$\nu_q$ can also be interpreted as the distribution of $\sum_{k=0}^{\infty} \sigma_k q^k$, where numbers $\sigma_k \in \{0, 1\}$ are chosen independently with probability $\frac{1}{2}$, $\nu_q$ is the infinite convolution product of $\frac{1}{2}(\delta_0 + \delta_{q^k})$. It also can be viewed as self-similar measure (see [Hutchinson (1981)]) for the IFS $\{qx, qx+1\}$ with probabilities $(\frac{1}{2}, \frac{1}{2})$.*

There are many interesting questions connected to this measure in harmonic analysis, algebraic numbers theory, dynamical systems, Hausdorff dimension estimation. The most famous is the question of Wintner and Erdős which stays open since 1935:

- Describe all values of $q$ for which this measure has a density function.

Jessen & Wintner [1935] showed that $\nu_q$ is either absolutely continuous, or purely singular, depending on $q$. Kershner & Wintner [1935] proved that $\nu_q$ is singular for $q \in (0, \frac{1}{2})$, Wintner [1935] found that $\nu_q$ is uniform on $[0, 2]$ for $q = \frac{1}{2}$. Finally, Erdősh [1939] showed that $\nu_q$ is singular when $q$ is the reciprocal of a Pisot number. Since that time and until now there are no more known values of $q$, for which $\nu_q$ is singular. Thorough discussion on this theme can be found in [Peres *et al.*, 2000].

## References


1. Bandt, C. [2002] "On the Mandelbrot set for pairs of linear maps," *Nonlinearity* **15**, 1127–1147.
2. Bandt, C. & Keller, K. [1991] "Self-similar sets 2. A simple approach to the topological structure of fractals," *Math. Nachr* **154**, 27–39.
3. Bandt, C. & Graf, S. [1992] "Self-similar sets 7. A characterization of self-similar fractals with positive Hausdorff measure," *Proc. Amer. Math. Soc* **114**, 995–1001.
4. Bandt, C. & Hung, N. V. [2008] "Self-similar sets with open set condition and great variety of overlaps," *Proc. Amer. Math. Soc* **136**, 3895–3903.
5. Bandt, C. & Rao, H. [2007] "Topology and separation of self-similar fractals in the plane," *Nonlinearity* **20**, 1463–1474.
6. Barnsley, M. F. [1988] *Fractals everywhere*, Academic Press, Boston.
7. Barnsley, M. F. [2005] "Theory and application of fractal tops," *Fractals in engineering: new trends in theory and applications*, eds. Lévy-Véhel, J. & Lutton, E, Springer-Verlag, London Limited, 3–20.
8. Barnsley, M. F. [2006] *Superfractals*, Cambridge University Press, Cambridge.
9. Barnsley, M. F. & Demko, S. [1985] "Iterated function systems and the global construction of fractals," *Proc. Roy. Soc. London* **399**, 243–275.
10. Barnsley, M. F. & Harrington, A. [1985] "A Mandelbrot set for pairs of linear maps," *Phis. D* **15**, 421–432.
11. Bousch, T. [1993] "Connexité locale et par chemins hölderiens pour les systèmes itérés de fonctions," `http://topo.math.u-psud.fr/~bousch/preprints/clh_ifs.pdf`.
12. Dube, S. [1994] "Undecidable problems in fractal geometry," *Complex systems: mechanism of adaptation*, eds. Stonier, R. J. & Yu, X. H. (IOS Press, Amsterdam), 283–290.
13. Erdős, P. [1939] "On a family of symmetric Bernoulli convolutions," *Amer. J. Math* **61**, 974–975.
14. Fisher, Y. (ed.) [1995] *Fractal image compression: theory and application*, Springer-Verlag, New York.
15. Hutchinson, J. E. [1981] "Fractals and self-similarity," *Indiana Unic. Math. J* **30**, 713–747.
16. Indlekofer, K. -H., Járai, A. & Kátai, I. [1995] "On some properties of at-

tractors generated by iterated function systems," *Acta Sci. Math. (Szeged)* **60**, 411–427.

17. Indlekofer, K. -H., Kátai, I. & Racsko, P. [1993] "Some remarks on generalized number systems," *Acta Sci. Math. (Szeged)* **57**, 543–555.
18. Jessen, B. & Wintner, A. [1935] "Distribution functions and the Riemann zeta function," *Trans. Amer. Math. Soc* **38**, 48–88.
19. Kechris, A. S. [1995] *Classical descriptive set theory*, Graduate texts in mathematics **156**, Springer-Verlag, New York.
20. Kershner, R. & Wintner, A. [1935] "On symmetric Bernoulli convolutions," *Amer. J. Math* **57**, 541–548.
21. Moran, P. A. P. [1946] "Additive functions of intervals and Hausdorff measure," *Proc. Cambridge Philos. Soc* 15–23.
22. Peres, Y. & Schlag, W. & Solomyak, B. [2000] "Sixty years of Bernoulli convolutions," *Fractal Geometry and Stochastics 2*, Progress in probability, Vol. **46**, eds. Bandt, C. & Graf, S. & Zaehle, M. (Birkhäuser, Basel), 39–65.
23. Rényi, A. [1957] "Representations for real numbers and their ergodic properties," *Acta Math. Acad. Sci. Hung* **8**, 477–493.
24. Solomyak, B. [1998] "Measure and dimension of some fractal families," *Math. Proc. Camb. Phil. Soc* **124**, 531–46.
25. Solomyak, B. [2005] "On the Mandelbrot set for pairs of linear maps: asymptotic self-similarity," *Nonlinearity* **18**, 1927–1943.
26. Solomyak, B. & Xu, H. [2003] "On the 'Mandelbrot set' for a pair of linear maps and complex Bernoulli convolutions," *Nonlinearity* **16**, 1733–1749.
27. Wintner, A. [1935] "On convergent Poisson convolutions," *Amer. J. Math* **57**, 827–838.

# Chapter 11

## Recent Advances in Open Billiards with Some Open Problems

Carl P. Dettmann

*School of Mathematics, University of Bristol, UK*
*E-mail: Carl.Dettmann@bristol.ac.uk*

Much recent interest has focused on "open" dynamical systems, in which a classical map or flow is considered only until the trajectory reaches a "hole", at which the dynamics is no longer considered. Here we consider questions pertaining to the survival probability as a function of time, given an initial measure on phase space. We focus on the case of billiard dynamics, namely that of a point particle moving with constant velocity except for mirror-like reflections at the boundary, and give a number of recent results, physical applications and open problems.

**Keywords**: Open dynamical systems, survival probability, mathematical billiards.

### Contents



### 11.1. Introduction

A mathematical billiard is a dynamical system in which a point particle moves with constant speed in a straight line in a compact domain $\mathcal{D} \subset \mathbb{R}^d$

with a piecewise smooth[a] boundary $\partial\mathcal{D}$ and making mirror-like[b] reflections whenever it reaches the boundary. We can assume that the speed and mass are both equal to unity. In some cases it is convenient to use periodic boundary conditions with obstacles in the interior, so $\mathbb{R}^d$ is replaced by the torus $\mathbb{T}^d$. Here we will mostly consider the planar case $d = 2$. Billiards are of interest in mathematics as examples of many different dynamical behaviors (see Fig.11. 1 below) and in physics as models in statistical mechanics and the classical limit of waves moving in homogeneous cavities; more details for both mathematics and physics are given below. An effort has been made to keep the discussion as non-technical and self-contained as possible; for further definitions and discussion, please see [Chernov & Markarian, 2006; Demers & Young, 2006; Gaspard, 1998; Gutkin, 2003]. It should also be noted that the references contain only a personal and very incomplete selection of the huge literature relevant to open billiards, and that further open problems may be found in the recent reviews [Bunimovich, 2008; Demers & Young, 2006; Nonnenmacher, 2008; Sinai, 2010; Szász, 2008].

The structure of this paper is as follows. In Secs. 11. 2 and 11. 3 we consider general work on closed and open dynamical systems respectively. In Sec. 11.4 we consider each of the main classes of billiard dynamics, and in Sec. 11.5 we consider physical applications. Sec. 11.6 returns to a more general discussion and outlook.

## 11.2. Closed Dynamical Systems

In this section, we introduce some notations for mathematical billiards, as well as informal descriptions of properties used to characterize chaos in billiards and more general systems, with pointers to more precise formulations in the literature. Much relevant work on open systems applies to chaotic maps different from billiards, so this notation is designed to be applicable both to billiards and to more general dynamical systems. A readable intro-

[a] The amount of smoothness required depends on the context; an early result of Lazutkin [Lazutkin, 1973] required 553 continuous derivatives, but recent theorems for chaotic billiards typically require three continuous derivatives except at a small set of singular points [Bunimovich & Grigo, 2010] while polygonal billiards and most explicit examples are piecewise analytic. For an attempt in a (non-smooth) fractal direction, see [Lapidus & Neimeyer, 2009].

[b] That is, the angle of incidence is equal to the angle of reflection. Popular alternative reflection laws include that of outer billiards, also called dual billiards, in which the dynamics is external to a convex domain, see for example [Dolgopyat & Fayad, 2009], and Andreev billiards used to model superconductors [Cserti *et al.*, 2009]; we do not consider these here.

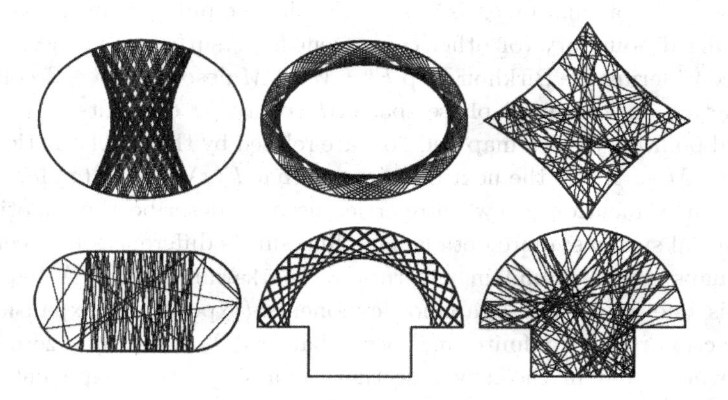

Fig. 11.1. Geometry vs. dynamics. An integrable elliptical billiard (top left and center) has two types of regular motion depending on the initial condition, a dispersing diamond (top right) is chaotic, a defocusing stadium (bottom left) is intermittent, with orbits switching between regular and chaotic motion, and a mushroom (bottom center and right) is mixed, with regular or chaotic motion depending on the initial condition. Apart from the ellipse, all these are constructed from circular arcs and straight lines; many other examples with more subtle dynamical distinctions exist. See Sec. 11.4 for discussion of all these cases.

duction to the subject of (closed) billiards is given in [Tabachnikov, 2005] with more details of chaotic billiards in [Chernov & Markarian, 2006].

Below, $|\cdot|$ will denote the size of a set, using the Lebesgue measure of the appropriate dimension.[c] We denote the dynamics by $\Phi^t : \Omega \to \Omega$ for either flows (including the case of billiards) or maps. For flows we have $t \in \mathbb{R}$ while for maps we have $t \in \mathbb{Z}$. $\Phi^t$ also naturally acts on subsets $A \subseteq \Omega$. For non-invertible maps and $t < 0$, action on points is undefined, but action on sets gives the relevant pre-image. The phase space $\Omega$ for the billiard flow consists of all[d] particle positions $q \in \mathcal{D}$ and momentum directions $p \in \mathbb{S}^{d-1}$ except that we need an extra condition to make $\Phi^t$ single-valued at collisions: For definiteness assume that $p$ points toward the interior of $\mathcal{D}$ if $q$ is at the boundary.

[c] All sets are assumed to be measurable with respect to the relevant measure(s).

[d] Billiard dynamics cannot usually be continued uniquely if the particle reaches a corner, so strictly speaking we need to exclude a zero measure set of points that do this at some time in the past and/or future. Another barrier to defining $\Phi^t$ is where there are infinitely many collisions in finite time; [Chernov & Markarian, 2006] states conditions under which this is impossible, roughly that for $d = 2$ there are finitely many corners, and focusing (convex) parts of the boundary are sufficiently ($C^3$) smooth, have non-zero curvature, and do not end in a cusp (zero angle corner). This is the case for almost all billiards considered in the past; exceptions may be interesting to consider in the future.

For many calculations it is more convenient to put a Poincaré section on the billiard boundary (or other convenient hypersurface for general flows), thus considering the Birkhoff map $F^n : M \to M$ describing $n \in \mathbb{Z}$ collisions, and acting on a reduced phase space $M$ consisting of points $x \in \partial\mathcal{D}$ and inward pointing $p$. The map and flow are related by the "roof function", the time $\tau : M \to \mathbb{R}^+$ to the next collision, so that $F(x) = \Phi^{\tau(x)}(x)$ for $x \in M$.

We now mention a few[e] properties used to describe the chaoticity of dynamical systems; more details including subtle differences between flows and maps can be found in [Chernov & Markarian, 2006]. A hyperbolic map is one where the Lyapunov exponents (exponential expansion and contraction rates of infinitesimal perturbations) $\lambda_i$ are all non-zero almost everywhere, and in the flow case there is a single zero exponent associated with the flow direction. For a uniformly hyperbolic system the relevant statement is true everywhere, with uniform bounds on the exponents and associated constants. For smooth uniformly hyperbolic (Anosov) systems the dynamics is controlled by unstable and stable manifolds in phase space corresponding to the positive and negative Lyapunov exponents respectively, yielding Sinai-Ruelle-Bowen (SRB) invariant measures smooth in the unstable directions, and dividing phase space into convenient entities called Markov partitions. Hyperbolicity that is nonuniform and/or non–smooth has similar properties, but requires more general and detailed techniques [Barreira & Pesin, 2007].

For ergodic properties we choose an invariant measure[f] $\mu$. For billiards there are natural "equilibrium" invariant probability measures $\mu_\Omega$ and $\mu_M$. $\mu_\Omega$ is given by the usual (Lebesgue) measure on the phase space $\Omega$, and $\mu_M$ is given by the product of measure on the boundary and the components of momentum parallel to the boundary. For example, in two dimensions $d\mu_M = ds\ dp_\parallel/(2|\partial\mathcal{D}|) = \cos\psi\ ds\ d\psi/(2|\partial\mathcal{D}|)$, where $s$ is arc length, $p_\parallel$ is momentum parallel to the boundary and $\psi$ is angle of incidence, that is, the angle between the momentum following a collision and the inward normal to the boundary. For integrable billiards such as the ellipse of Fig. 11.1, this is one of many smooth invariant measures defined by arbitrary smooth functions on the level sets of the conserved quantity, while for ergodic billiards such as the stadium or diamond of Fig. 11.1, it is the only smooth invariant measure.

---

[e]The properties mentioned are those commonly discussed in the literature of open dynamical systems including billiards, and tend to be measure theoretic rather than topological.

[f]That is, $\mu(\Phi^{-t}A) = \mu(A)$ for any $A \subseteq \Omega$ and any $t$ (positive for non–invertible maps).

Given an invariant measure $\mu$, recurrence is the statement that $\mu$-almost all trajectories return arbitrarily close to their initial point; it is guaranteed by the Poincaré recurrence theorem if $\mu$ is finite as above. Ergodicity is the statement that all invariant sets have zero or full measure, which implies that time averages of an integrable phase function are for almost all (with respect to $\mu$) initial conditions given by its phase space average over $\mu$. Mixing is the statement that the $\mu$-probability of visiting region $A$ at time zero and $B$ at time $t$ are statistically independent in the limit $|t| \to \infty$ for sets $A, B \subset \Omega$. There is an equivalent statement in terms of decay of correlation functions such as found in Eq. (11.9). Stronger ergodic properties are Kolmogorov mixing (K-mixing) and Bernoulli. We have Bernoulli $\Rightarrow$ K-mixing $\Rightarrow$ mixing $\Rightarrow$ ergodicity $\Rightarrow$ recurrence.

Further statistical properties build on the above. These include rates of decay of correlation functions for mixing systems and moderately regular (typically Hölder continuous) phase functions and central limit theorems for time averages of phase functions. See [Bálint & Melbourne, 2008; Barreira & Pesin, 2007; Chernov & Zhang, 2009] for a discussion of recent results in this direction.

Finally, the Kolmogorov-Sinai (KS) entropy $h_{KS}$ is a measure of the unpredictability of the system. For a closed, sufficiently smooth map with a smooth invariant measure it is equal to the sum of the positive Lyapunov exponents (if any); this is called Pesin's formula, and has been proved for some systems with singularities including billiards. All K-mixing systems have $h_{KS} > 0$.

Note that many of these descriptors of chaos are logically independent. In the case of billiards, open problem 11.6 conjectures that polygonal billiards may be mixing, but they are certainly not hyperbolic, having zero Lyapunov exponents everywhere. The recently proposed "track billiards" [Bunimovich & Del Magno, 2009] are hyperbolic but not ergodic or mixing. The stadium of Fig. 11.1 and the section below on defocusing billiards is Bernoulli [Chernov & Markarian, 2006] but has slow decay of correlations [Barreira & Pesin, 2007] and a non-standard central limit theorem [Bálint & Gouësel, 2006].

## 11.3. Open Dynamical Systems

Open dynamical systems are reviewed in both the mathematical [Demers & Young, 2006] and physical [Altmann & Tél, 2009] literature. Most of the mathematical studies of open dynamical systems have considered strongly chaotic systems, such as piecewise expanding maps of the interval or Anosov (uniformly hyperbolic) maps in higher dimensions. Note that there are a variety of conventions used in the literature to describe open systems.

An open dynamical system contains a hole $H \subset \Omega$ at which the particle is absorbed and no longer considered. $H$ may have more than one connected component (several "holes"), and may be on the boundary (ie a subset of $M$) or in the interior, but should be piecewise smooth and of dimension one less than $\Omega$ for flows, and of the same dimension as $\Omega$ for maps. Here we allow a particle to be injected at the hole, so it is absorbed only when it reaches the hole at strictly positive time. If we denote by $\Omega_t$ the subset of $\Omega$ that does not reach the hole by time $t$,

$$\Omega_t = \{x \in \Omega : \Phi^s(x) \notin H, \quad \forall s \in (0, t]\} \tag{11.1}$$

a typical question to ask is that given a set of initial conditions distributed according to some probability measure $\mu_0$ at time $t = 0$, what is the probability $P(t) = \mu_0(\Omega_t)$ that the particle survives until time $t$? How does this probability behave as a function of $t$, the initial measure $\mu_0$, the hole location $H$, its size relative to the billiard boundary $h = |H|/|M|$ (for general maps, $|H|/|\Omega|$), and the shape of the billiard $\mathcal{D}$?

In mathematical literature [Lopes & Markarian, 1996; Petkov & Stoyanov, 2009] the term "open billiard" has been used to incorporate additional conditions. In terms of our notation the outer boundary of $\mathcal{D}$ is a strictly convex set forming the "hole" through which particles escape; $\mathcal{D}$ also excludes three or more strictly convex connected obstacles in its interior satisfying a non-eclipsing condition, that is, the convex hull of the union of any two obstacles does not intersect any other obstacles. This ensures that there is a trapped set of orbits that never escape, hyperbolic and with a Markov partition, leading to a relatively good understanding [Eckhardt *et al.*, 1994; Gaspard, 1998; Lopes & Markarian, 1996; Petkov & Stoyanov, 2009; Sjöstrand, 1990]. Billiards satisfying these conditions will be denoted here as "non-eclipsing" billiards.

For strongly chaotic systems (including the Anosov maps mentioned above and dispersing billiards with finite horizon discussed below) we expect that $P(t)$ decays with time as described by an (exponential) escape rate

$$\gamma = -\lim_{t\to\infty} \frac{1}{t} \ln P(t) \tag{11.2}$$

which exists and is independent of a reasonable class of initial measures $\mu_0$.[g] We recall that $t$ corresponds either to discrete or continuous time depending on the context, with (the more physical) continuous time implied for billiards except where otherwise stated.

We now discuss conditionally invariant measures; see also [Demers & Young, 2006]. The renormalized evolution $\Phi^t_H$ of the initial measure $\mu_0$ is defined by its action on sets $A \subseteq \Omega$:

$$\mu_t(A) = (\Phi^t_H \mu_0)(A) = \frac{\mu_0(\Phi^{-t}(A) \cap \Omega_t)}{\mu_0(\Omega_t)} \tag{11.3}$$

It is easy to check that $\Phi^t_H$ satisfies the semigroup property $\Phi^s_H \circ \Phi^t_H = \Phi^{s+t}_H$ for non-negative $s$ and $t$. If $\mu_t$ approaches a limit $\mu_\infty$ at which $\Phi^t_H$ is continuous (in a suitable topology), this limit is conditionally invariant, ie $\Phi^t_H \mu_\infty = \mu_\infty$ for all $t \geq 0$, and gives the escape rate

$$\mu_\infty(\Omega_t) = e^{-\gamma t} \tag{11.4}$$

also for all $t \geq 0$.

[Kaufmann & Lustfeld, 2001] makes the important point that for open systems, conditionally invariant measures of $\Phi$ (projected onto $M$) and $F$, while supported on the same set, are generally not equivalent measures, even when the roof function $\tau$ is smooth, in contrast to the closed case. In other words escape properties of the flow do not follow trivially from those of the map. For billiard calculations, it is usually best to work first with the collision map $F$ (projecting any interior holes onto the collision space $M$), then try to incoporate effects of the flow. For incorporation of flow effects in proofs of statistical properties of closed billiards, see [Barreira & Pesin, 2007; Chernov & Markarian, 2006; Bálint and Melbourne 2008] and in calculation of escape rates and averages of open systems, see [Bunimovich & Dettmann, 2007; Gaspard, 1998; Lan, 2010].

Discussion in [Demers & Young, 2006] demonstrates that conditionally invariant measures can have properties similar to the SRB measures of closed hyperbolic systems, for example being smooth in the unstable direction. For non-invertible maps the measures satisfying the property of conditional invariance can however be highly non-unique.

The conditionally invariant measure $\mu_\infty$ is supported on the set of points with infinite past avoiding the hole, however most of these points reach the

[g]In [Demers & Young, 2006] some examples are given where existence is shown for $\mu_0$ equivalent to Lebesgue with density bounded away from 0 and $\infty$; this condition is probably too restrictive.

hole in the future. There is an invariant set of points never reaching the hole in past or future called the repeller; the natural invariant measure on this set is defined by the limit

$$\nu(A) = \lim_{t\to\infty} e^{\gamma t}\mu_\infty(A \cap \Omega_t) \tag{11.5}$$

Pesin's formula generalizes in the open case to the "escape rate formula" [Demers & Young, 2006; Gaspard, 1998]

$$\gamma = \sum_{i:\lambda_i>0} \lambda_i - h_{KS} \tag{11.6}$$

where the quantities on the right hand side are defined with respect to $\nu$.[h] The escape rate formula has very recently been shown for a system that is not uniformly hyperbolic [Bruin *et al.*, 2010], however there is wide scope for further development both in maps without uniform hyperbolicity and in flows.

**Problem 11.1.** *How general is the escape rate formula, Eq. (11.6)?*

There has been a very recent explosion of interest in the mathematics of open dynamical systems [Bunimovich & Yurchenko, 2010; Demers *et al.*, 2010; Keller & Liverani, 2009; Petkov & Stoyanov, 2009], again mostly restricted to strongly hyperbolic maps as above. In [Bunimovich & Yurchenko, 2010; Keller & Liverani, 2009] it is shown that $\gamma/h$ can reach a limit, the local escape rate, as the hole shrinks to a point, and that the limit depends on whether the point is periodic. For example consider the map $f(x) = 2x \pmod 1$ and a point $z$ of minimal period $p$ so that $f^p(z) = z$, together with a sequence of holes $H_n$ of size $h_n$ and each containing $z$, with corresponding escape rates $\gamma_n$. Then there is a local escape rate

$$\lambda_z = \lim_{n\to\infty} \frac{\gamma_n}{h_n} = 1 - 2^{-p} \tag{11.7}$$

If the point $z$ is aperiodic, then $\lambda_z$ is equal to 1. For more general 1D maps, the $2^{-p}$ is replaced by the inverse of the stability factor $|(d/dx)f^p(x)|$ at $x = z$ and the invariant measure of the map (uniform in the above example) needs to be taken into account.

This and related results are, however available only for piecewise expanding maps and closely related systems, although "it is conceivable that

[h] Proofs of the escape rate formula typically replace the sum of Lyapunov exponents by the (more general) Jacobian of the dynamics restricted to unstable directions. For two dimensional billiards, at most one Lyapunov exponent is positive.

in the near future this result could be applied e.g. to billiards." [Keller & Liverani, 2009]. The very recent work [Demers *et al.*, 2010] on the periodic Lorentz gas, a dispersing billiard model (see below) shows that $\gamma$ (defined with respect to collisions, not continuous time) is well defined for sufficiently small holes, corresponding to a limiting conditionally invariant measure that is independent of the initial measure for a relatively large class of the latter, also that $\gamma \to 0$ as $h \to 0$, but does not say anything about $\gamma/h$. This leads to the open problem:

**Problem 11.2.** *Local escape rate: Proof of a formula similar to Eq. (11.7) for sufficiently chaotic billiard models.*

Note that the holes typically considered in billiards, consisting of a set small in position but with arbitrary momentum,[i] include phase space distant from any short periodic orbit contained in the hole. Following the discussion in [Bunimovich & Dettmann, 2007], the effects of the periodic orbit may then appear at a higher power of the hole size, compared to the above piecewise expanding map, so for example the formula might look something like $\gamma = h + h^2\gamma_2 + O(h^3)$ with $\gamma_2$ depending on properties of the shortest periodic orbit contained in the hole. Also note (see open problem 11.4 below) that if there is a local escape rate, $P$ behaves like $e^{-\lambda_z ht}$, in other words like a function of the combination $ht$ in this limit. Another important question posed in this work is that of optimization: how to choose a hole position to minimize or maximize escape, for example $\gamma$ or even the whole function $P(t)$ [Afraimovich & Bunimovich, 2010; Bunimovich & Yurchenko, 2010].

**Problem 11.3.** *Optimization: Specify where to place a hole to maximize or minimize a suitably defined measure of escape.*

In the above papers, slow escape is related to placing the hole on a short periodic orbit or on the part of phase space with the lowest "network load." The main question is how these criteria generalize to open dynamical systems (including billiards) with distortion, nonuniform hyperbolicity or no hyperbolicity.

As will become clear below, the existence of an exponential escape rate $\gamma$ and nontrivial conditionally invariant measure $\mu_\infty$ is only expected in

[i]Note that other possibilities occur in applications (Sec.11.5): Escape for trajectories with sufficiently small angle of incidence but at any boundary point is relevant to microlasers, and escape with probability depending on the boundary material is relevant to room acoustics.

fairly special cases: there are many systems including billiards with slower (or occasionally faster) decay of the survival probability. For some systems with slower eventual decay of $P(t)$, a cross-over from approximately exponential decay at short times to algebraic decay at late times has been observed numerically [Altmann & Tél, 2009]. We now turn to our main focus, that of open billiards, returning to general open dynamical systems in the applications section.

## 11.4. Open Billiards

Now we consider open billiards, categorized according to dynamical properties of the corresponding closed case. The initial measure $\mu_0$ for an open billiard is normally given by the equilibrium measure for the flow or Poincaré map, $\mu_\Omega$ or $\mu_M$ respectively. An alternative is to consider injection at the hole by a restriction of the equilibrium measure, in other words transport through the billiard; in many cases this has little effect on the long time properties [Altmann & Tél, 2009]. A few papers discussed here and in the statistical mechanics section below consider billiards on infinite domains, usually consisting of an infinite collection of non-overlapping obstacles. If both the billiard and any hole(s) are periodic, this is equivalent to motion on a torus; an example is [Demers *et al.*, 2010]. However, if the billiard is aperiodic, or the hole is not repeated periodically, the above definitions break down because the equilibrium billiard measure is not normalizable. The above definition of ergodicity makes sense using an infinite invariant measure, but other properties including mixing and survival probability are problematic. Also, the Poincaré recurrence theorem fails and so recurrence properties need to be proved. One approach to defining $P(t)$ in the case of a non-repeated hole might be to choose an initial measure $\mu_0$ with support in a bounded region, somewhat analogous to the finite (but rather arbitrary) outer boundary of the domain $\mathcal{D}$ chosen for non-eclipsing billiards above.

**Integrable billiards:** A few billiards are integrable, i.e., perfectly regular, namely the ellipse, rectangle, equilateral triangle and related cases [Tabachnikov, 2005]. In the integrable circle with a hole of angle $2\pi h$ in the boundary [Bunimovich & Dettmann, 2005], the leading order coefficient of $P(t)$ was obtained exactly; in particular the statement

$$\lim_{h\to 0}\lim_{t\to\infty} h^{\delta-1/2}(tP(t)-\frac{1}{\pi h})=0 \qquad (11.8)$$

for all $\delta > 0$ is equivalent to the Riemann Hypothesis, the greatest unsolved problem in number theory.[j] If instead both limits are taken simultaneously so that $ht$ is constant, the survival probability numerically appears to reach a limiting function.

**Problem 11.4.** *Scaling: For what billiards and in what limits does $P(t)$ reduce to a function of $ht$ for a family of billiards with variable hole size $h$?*

**Polygonal billiards:** The survey [Gutkin, 1996] gives a good introduction to polygonal billiards. A billiard collision involving a straight piece of boundary is equivalent to free motion in an "unfolded" billiard reflected across the boundary as in a mirror, using multiple Riemann sheets if the reflection leads to an overlap. General polygonal billiards with angles rational multiples of $\pi$ can be unfolded into flat manifolds with nontrivial topology and conical singularities (from the corners), called translation surfaces, which are currently an active area of interest [Athreya & Forni, 2008]. In such billiards each trajectory explores only a finite number of directions, however with this restriction is typically ergodic but not mixing [Ulcigrai, 2009]. There has been recent progress on showing recurrence for infinite rational polygonal billiards [Troubeztkoy, 2010; Gutkin, 2010].

Very little is known about the case of irrational angles [Jepps *et al.*, 2008; Valdez, 2009], except that usual characteristics of chaoticity such as positive Lyapunov exponents are not possible here. We have the well known open question [Schwartz, 2009]:

**Problem 11.5.** *Existence of periodic orbits: Do all triangular (or more generally polygonal) billiards contain at least one periodic orbit?*

Note that this admits arbitrary directions; it is easy to construct rational polygonal billiards for which a particular directional flow is never periodic. Also, there are a number of cases including rational and acute triangles for which periodic orbits can be explicitly constructed. Periodic orbits in polyhedra are considered in [Bedaride, 2008].

Another open problem has the interesting feature that mathematical literature conjectures that it is false [Gutkin, 1996], while physical literature conjectures that it is true [Casati & Prosen, 1999]; recent work is found in [Ulcigrai, 2009]:

---

[j] The Riemann hypothesis is the statement that all the complex zeros of the Riemann zeta function, the analytic continuation of $\zeta(s) = \sum_{n=1}^{\infty} n^{-s}$ lie on the line $Re(s) = 1/2$, and is related to the distribution of prime numbers [Conrey, 2003].

**Problem 11.6.** *Mixing property: Is it possible for a polygonal billiard (necessarily with at least one irrational angle) to be mixing?*

There is very little work on open polygonal billiards given the paucity of knowledge about the closed case. In [Dettmann & Cohen, 2000] it was found numerically that $P(t)$ decayed as $C/t$ in an irrational polygonal billiard where at least one periodic orbit was present, and vanished at finite time in a directional flow for a rational polygonal billiard containing no periodic orbits. So we have

**Problem 11.7.** *Is the asymptotic survival probability $P(t)$ as $t \to \infty$ in polygonal billiards entirely determined by the neighbourhoods of non-escaping periodic orbits?*

**Dispersing billiards:** Curved boundaries of billiards lead to focusing (convergence) or dispersion (divergence) of an initially parallel set of incoming trajectories, depending on the sign of the curvature.[k] [l] We have already noted that one focusing billiard, the ellipse, is integrable; more generally all sufficiently smooth strictly convex billiards are not ergodic [Lazutkin, 1973]. However there are a number of other classes of billiards which are not only ergodic, but have much stronger chaotic properties, to which we turn; [Chernov & Markarian, 2006] provides a good introduction and [Sinai, 2010; Szasz, 2008] recent reviews.

Beginning with the Sinai billiard [Sinai, 1970] which used a convex obstacle in a torus to ensure that all collisions were dispersing, there have been many new constructions of chaotic billiards and techniques to prove stronger statistical properties of existing billiards. For smooth dispersing cases such as Sinai's billiard, ergodic properties now include the Bernoulli property. The theory is technical due to singularities in the dynamics, requiring additional techniques for more difficult classes of singularity: All dispersing billiards have trajectories tangent to one or more parts of the boundary; perturbed trajectories either miss that part of the boundary entirely, or collide with an outgoing perturbation proportional to the square

[k] Zero curvature, ie inflection points and similar, can be problematic; see footnote d. The main exception is where part of the boundary is exactly flat. A class of billiards allowing some inflection points that has been studied in detail is that of semi-dispersing billiards [Chernov & Simányi,2010, Szász, 2008].

[l] In three or more dimensions the curvature at a boundary point generally depends on the plane of the incoming trajectory; this phenomenon is known as astigmatism and causes difficulties in the theory, hence many results are weaker and/or require additional assumptions; see for example [Bálint & Tóth, 2008, Petkov & Stoyanov, 2009].

root of the incoming one, that is, infinite linear instability. In addition, the boundary may contain corners, in most cases leading to discontinuities in the dynamics. In the case of cusps (zero angle corners) the number of collisions near the cusp in a finite time is unbounded. A further complication occurs in motion on a torus if some trajectories never collide ("infinite horizon condition"),[m] due to an accumulation of singularity sets near such orbits. Proof of exponential decay of correlations of the billiard map was possible using Young tower constructions of the late 1990s which generalize the concept of a Markov partition, but exponential decay of correlations for the flow is still conjectured; recent results for advanced statistical properties and treatment of continuous time may be found in [Barreira & Pesin, 2007; Chernov & Zhang, 2009] and some open problems in [Szász, 2008].

An example of a dispersing billiard with corners is the diamond shown in Fig. 11.1. This was the numerical example in [Bunimovich & Dettmann, 2007], where the exponential escape rate $\gamma$ as above can be expanded in powers of the hole size $h$ using series of correlation functions, as can the differential escape rate of one vs two holes. For example, the expression for the latter, following a derivation that is far from rigorous but consistent with numerical tests, is

$$\gamma_{AB} = \gamma_A + \gamma_B - \frac{1}{\langle\tau\rangle} \sum_{j=-\infty}^{\infty} \langle (u_A)(F^j \circ u_B) \rangle + \dots \tag{11.9}$$

where $A$ and $B$ label the holes, $\tau$ is the time from one collision to the next, $u$ is a zero mean phase function (that is, $\langle u \rangle = 0$) equal to $-1$ on the relevant hole and $h\tau/\langle\tau\rangle$ elsewhere, angle brackets indicate integration with respect to normalized equilibrium measure on the billiard boundary $\mu_M$, and $F$ is the billiard map. The omitted terms have third or higher order correlations, and are expected to be of order $h^3$ as long as $A$ and $B$ do not contain points from the same short periodic orbit. Convergence of the sum over $j$ follows from sufficiently fast decay of correlations, which do not need to be exponential. An extended diamond geometry was used as a model for heat conduction in [Gaspard & Gilbert, 2008].

**Problem 11.8.** *Give a rigorous formulation of the escape rate of dispersing billiards with small holes along the lines of Eq. (11.9).*

---

[m] In the context of infinite billiards, there has recently been proposed a "locally finite horizon condition", in which all straight lines pass through infinitely many scatterers, but the free path length is unbounded, see [Troubeztkoy, 2010]. Here it is also assumed that the minimum distance between obstacles be bounded away from zero.

Dispersing billiards, which include the non-eclipsing billiards defined above, also provide a likely setting for solving the open problems in Sec. 11.2.

**Defocusing and intermittent billiards:** In addition to the dispersing billiards, strongly chaotic billiards may be constructed with the defocusing mechanism of Bunimovich, in which trajectories that leave a focusing piece of boundary have a sufficient time to defocus (ie disperse) before reaching another curved part of the boundary; for a recent discussion see [Bunimovich & Grigo, 2010]. In this case there are difficulties due to "whispering gallery" orbits close to a focusing boundary with finite but unbounded collisions in a finite time. Some concepts and results can be derived from the Pesin theory of non–uniformly hyperbolic dynamics [Barreira & Pesin, 2007; Chernov & Markarian, 2006].

An example of a defocusing billiard is the stadium of Fig. 11.1; note that while it is ergodic and hyperbolic, the presence of "bouncing ball" orbits between the straight segments leads to intermittent quasi-regular behavior, leading to some weaker statistical properties, so for example while the system is mixing, decay of correlations is now $C/t$ for both the map and flow, leading to a non-standard central limit theorem with $\sqrt{n \ln n}$ (rather than $\sqrt{n}$) normalization [Bálint & Gouëzel, 2006; Barreira & Pesin, 2007].

The open stadium has been considered in [Armstead *et al.*, 2004; Dettmann & Georgiou, 2009]; this is a good model for intermittency due to "bouncing ball" motions between the straight segments. The first paper discusses scaling behavior associated with evolution of measures at long times and small angles close to the bouncing ball orbits, while the second finds an explicit expression for the leading coefficient of the survival probability

$$\lim_{t\to\infty} tP(t) = \frac{(3\ln 3 + 4)((a+h_1)^2 + (a-h_2)^2)}{4(4a+2\pi r)} \tag{11.10}$$

for a stadium with horizontal straight segments $x \in (-a, a)$ containing a small hole for $x \in (h_1, h_2)$ with $-a < h_1 < h_2 < a$, and semicircles of radius $r$. Note that this approaches a constant as $h \sim h_2 - h_1 \to 0$, thus demonstrating an example where fixed $ht$ is *not* a correct scaling limit (compare open problem 11.4).

The stadium is the most famous defocusing billiard, but many of its properties are due to the intermittency arising from the bouncing ball orbits, or more generally a family of marginally unstable periodic orbits, rather than the defocusing mechanism per se; defocusing billiards need not have such orbits [Bunimovich & Grigo, 2010]. One source of interest and also difficulty with the stadium is the fact that orbits that leave the bouncing ball region are immediately reinjected back, with an angle that is approximately described using an independent stochastic process [Armstead, *et al.*, 2004]. Other reinjection mechanisms are possible, depending on the properties of the curvature of the boundary approaching the end points of the bouncing ball orbits. In some cases small changes of the boundary of a stadium lead to a breakdown of ergodicity [Grigo, 2009], which of course will also affect relevant escape problems.

**Problem 11.9.** *Give a comprehensive characterization of the dynamical (including escape) properties of marginally unstable periodic orbits in stadium-like billiards in terms of their reinjection dynamics.*

**Mixed billiards:** Typical billiards are expected to have mixed phase space, that is, chaotic or regular depending on the initial condition, in a fractal hierarchy according to Kolmogorov-Arnold-Moser (KAM) theory [Broer, 2004]. Much remains to be understood about this generic case; it has been conjectured [Altmann & Tél, 2009; Cristadoro & Ketzmerick, 2008] that in the case of area preserving maps the long time survival probability associated with stickiness near the elliptic islands decays as $t^{-\alpha}$ with a universal $\alpha \approx 2.57$.

**Problem 11.10.** *Is there a universal decay rate in generic open billiards?*

One fruitful line of enquiry has been the introduction of mushroom billiards by Bunimovich; these have mixed phase space, but with a smooth boundary between the regular and chaotic regions. See Fig. 11.1 and [Bunimovich, 2008]. The chaotic region of most of these is intermittent ("sticky") due to embedded marginally unstable periodic orbits [Andreasen, *et al.*, 2009], however an example of a non-sticky mushroom-like billiard was given in [Bunimovich, 2008]. The decay of the survival probability in the case that the hole is in the chaotic region may be significantly slowed both by stickiness due to marginally unstable periodic orbits, and by the boundary itself.[n]

**Problem 11.11.** *Characterize the escape properties of the boundary between regular and chaotic regions in mushroom and similar billiards.*

[n]Recent unpublished result of the author in collaboration with O. Georgiou.

## 11.5. Physical Applications

**Statistical mechanics:** One important application of billiards is that of atomic and molecular interactions, for which steep repulsive potential energy functions can be approximated by the hard collisions of billiards. A system of many particles undergoing hard collisions corresponds to a high dimensional billiard. This type of billiard is described as semi-dispersing since during a collision of two particles the other particles are not affected, so there are many directions in the high dimensional collision space with zero curvature; however there has been recent progress in demonstrating ergodicity (conjectured by Boltzmann); see [Chernov & Simányi, 2010; Szász, 2008]. Note, however, that ergodicity may be broken by arbitrarily steep potential energies approximating the billiard [Rapoport & Rom-Kedar, 2008]:

**Problem 11.12.** *For physical systems with many particles which are predominantly chaotic, how prevalent and important are the regular regions ("elliptic islands") in phase space? [Broer, 2004, Bunimovich, 2008, Rapoport & Rom-Kedar, 2008].*

Systems of many hard particles have also been instrumental in the discovery of a fascinating connection between microscopic and macroscopic dynamical effects, that of Lyapunov modes, although later found in systems with soft potentials; for a review see [Yang & Radons, 2009]. There are many other connections between dynamics in general and statistical mechanics [Dettmann, 2000; Marconi *et al.*, 2008].

Low-dimensional billiards imitating hard-ball motion are popular for understanding statistical mechanics, particularly Lorentz gases consisting of an infinite number of convex (typically circular) scatterers and the Ehrenfest gases consisting of an infinite number of polygons [Dettmann, 2000; Gaspard, 1998; Marklof & Strömbergsson, 2010]. Infinite periodic arrays in these models are equivalent to motion on a torus and can be treated using similar techniques to those of equivalent finite billiards [Demers *et al.*, 2010; Gaspard & Gilbert, 2008]. However, very little has been proven about the more physically realistic models with randomly placed non-overlapping obstacles [Troubeztkoy, 2010]. There is numerical evidence for some statistical properties corresponding to a limiting Weiner (diffusion) process in some Ehrenfest gases [Dettmann & Cohen; 2001, Jepps *et al.*, 2008].

**Problem 11.13.** *What are minimum dynamical properties required for a chaotic macroscopic limit, for example recurrence, ergodicity, a normal diffusion coefficient, a limiting Weiner process?*

While most of the above work pertains to closed systems (albeit on large or infinite domains), it is worth pointing out that the "escape rate formalism" [Dettmann, 2000; Gaspard, 1998] relates escape rates in large open chaotic systems to linear transport properties; note that the escape rate is also related to other dynamical properties by Eq. (11.6). A final problem for this topic:

**Problem 11.14.** *What experimental techniques might best probe the limits of statistical mechanics arising from the previous two open problems?*

**Quantum chaos:** Billiards correspond to the classical (short wavelength) limit of wave equations for light, sound or quantum particles in a homogeneous cavity. The classical dynamics corresponds to the small wavelength, geometrical optics approximation. Semiclassical theory uses properties of this classical dynamics, especially periodic orbits, as the basis for a systematic treatment of the wave properties (eigenvalues and eigenfunctions of the linear wave operator). Comparisons are also made with the predictions of random matrix theory, in which the spacing of eigenvalues of many quantum systems follows universal laws based only on the chaoticity and symmetries of the problem [Poli *et al.*, 2009].[o] Billiards are both necessary for calculating properties of wave systems and a testbed for semiclassical and random matrix theories. A useful survey of quantum chaos, including open quantum systems, is given in [Nonnenmacher, 2008]; note that semiclassical approaches are also relevant to non-chaotic systems such as polygonal billiards [Hassell, *et al.*, 2009].

An important recent development in open quantum systems has been the fractal Weyl conjecture relating the number of resonances up to a particular energy to fractal dimensions of classically trapped sets. Following [Lu *et al.*, 2003] we recall that a quantum billiard is equivalent to the Helmholtz equation $(\nabla^2 + k^2)\psi = 0$ with a Dirichlet condition $\psi = 0$ at the boundary, the eigenvalues $k$ are labelled $k_n$ and are real. In this case Weyl's law gives a detailed statement to the effect that the number of states with $k_n < k$ is proportional to $Vk^d$ in the limit $k \to \infty$ where $d$ is the dimension of the space; effectively this means that each quantum state occupies the same classical phase space volume.

If the billiard is open, for example consisting of a finite region containing obstacles that do not prevent the particle escaping to infinity, the Helmholtz

[o]There are fascinating conjectures [Conrey *et al.*, 2008] relating random matrix theory and the Riemann zeta function of footnote j.

equation has resonance solutions where $k$ has negative imaginary part. We have [Lu *et al.*, 2003]:

**Problem 11.15.** *The fractal Weyl conjecture: The number of states with* $Re(k_n) < k$ *and* $Im(k_n) > -C$ *for some positive constant* $C$ *is of order* $k^{d_H+1}$ *where again* $k \to \infty$ *and* $d_H$ *is the partial Hausdorff dimension of the non-escaping set.*

While open billiards of the non-eclipsing type have been under investigation from the beginning [Sjöstrand, 1990], the most progress to date has been in simpler systems such as quantum Baker maps [Keating *et al.*, 2008; Novaes *et al.*, 2009]. Recent work has also included smooth Hamiltonian systems [Ramilowski *et al.*, 2009] and optical billiards [Wiersig & Main, 2008].

**Experiments and further applications:** The case of microresonators where light is trapped by total internal reflection is of interest for practical applications in laser design; the hole then corresponds to a condition on momentum rather than position, and desired properties typically include large $Q$-factor (small imaginary part of $k$) so that the pumping energy required for laser action is small, together with a strongly directional wave function at infinity. The earliest example of a circular cavity has large $Q$-factor due to trapping of light by total internal reflection, but the symmetry precludes a directional output, so a number of efforts have been made to modify the circular geometry, using insight from classical billiards [Andreasen, *et al.*, 2009; Dettmann, *et al.*, 2009; Pollinger, *et al.*, 2009].

Other experiments of relevance have included microwaves, sound, atoms and electrons in cavities at scales ranging from microns to metres. These are aimed at demonstrating the capabilities of new experimental techniques, testing theoretical results from quantum chaos, and laying the groundwork for specific applications. Often modifications to the original billiard problem arise. For example, open semiconductor billiards can be constructed by confining a two dimensional electron gas (2DEG) using electric fields. In transport through such cavities, a weak applied magnetic field shifts the quantum phases (hence conductance) while the classical orbits are effectively unchanged, while a strong field also curves the classical orbits; also the walls are likely to be described by a somewhat soft potential energy function. For room acoustics, collisions involve a proportion of the sound escaping, being absorbed, being randomly scattered, and being reflected using the usual billiard law, depending on the materials at the relevant point on the boundary. Relevant references may be found in [Bruin, *et al.*,

2010; Dettmann & Georgiou, 2009; Heller, 2008; Weaver & Wright, 2010]. Open dynamical systems that are more general than billiards are important for escape and transport problems in heat conduction [Gaspard & Gilbert, 2008], chemical reactions [Ezra, *et al.*, 2009], astronomy [Waalkens, *et al.*, 2005] and nanotubes [Jepps, *et al.*, 2008]. In these systems, open billiards provide a useful starting point for an understanding of more general classes of open dynamical systems.

**Problem 11.16.** *Generalizations: Billiards with more realistic physical effects, for example soft wall potentials, external fields, dissipative and/or stochastic scattering (at the boundary or in the interior) or time dependent boundaries.*

The possibilities are endless; for this reason, it is increasingly important to carefully justify the mathematical and/or physical interest of any new model.

## 11.6. Discussion

We conclude with a discussion of some problems that draw together many classes of dynamics.

**Problem 11.17.** *Exact expansions: For a given open (generalized) billiard problem, can the survival probability $P(t)$ be expressed exactly, or at least as an expansion for large $t$ that goes beyond the leading term?*

The most likely candidates for this problem are those billiards that are best understood, the integrable and dispersing cases.

One feature that is common to all classes of billiards, classical and quantum is the role of periodic orbits in the open problem. Periodic orbits denote an exact recurrence to the initial state of a classical system, and so contribute directly to the set of orbits that enter from a hole and return there after one period. Periodic orbits may be of measure zero, but some positive measure neighbourhood of a periodic orbit will be sufficiently close as to have the same property. For (finite) billiards with no periodic orbits the Poincaré recurrence theorem guarantees similar behaviour.

For an initial measure supported other than on the hole, periodic orbits are relevant to the set of orbits which never escapes, and its neighbourhood often determines the long time survival probability. This is seen in many of the above sections, for integrable, other polygonal, chaotic and intermittent billiards. There is substantial existing theory for calculating the escape rate $\gamma$ and other long time statistical properties (averages etc.) for hyperbolic systems, using either periodic orbits avoiding the hole [Lan, 2010] or passing through the hole [Altmann & Tèl, 2009], and also for semiclassical treatments of quantum systems [Nonnenmacher, 2008].

**Problem 11.18.** *Recurrence: Give a general description of how the escape problem is affected by recurrences in the system (periodic orbits or more general), in the vicinity of the hole and/or elsewhere, for systems with little or no hyperbolicity.*

Finally, a question posed in [Bunimovich & Dettmann, 2005] but for which a systematic solution does not appear to exist in the literature:

**Problem 11.19.** *Inverse problems: What information can be extracted about the dynamics of a billiard from escape measurements? Can you "peep" the shape of a drum? [Bunimovich & Dettmann, 2007, Kac, 1966].*

The open problems we have considered vary from the long standing and difficult (problem 11.5) to those arising very recently from active research that may as quickly solve them (problem 11.2). Even the simplest of integrable systems, the circle, has unexpected complexity, while a rigorous approach to hyperbolic billiards has a highly developed and technical theory, and the study of polygons with irrational angles is in its infancy. It is likely that progress will be made almost immediately on many of the problems in specific cases but that constructing reasonably complete solutions in general dynamical contexts will be a active area of research for many years. Finally, open billiards are interesting from a mathematical point of view and also required for solving practical problems; this would appear to be a particularly fruitful field for collaborations between mathematicians and physicists.

**Acknowledgements**

The author owes much to patient collaborators and other colleagues including Eduardo Altmann, Peter Bálint, Leonid Bunimovich, Eddie Cohen, Orestis Georgiou, Alex Gorodnik, Jon Keating, Edson Leonel, Jens Marklof and Martin Sieber for discussions on many topics related to this article.

## References


1. Afraimovich, V. S. & Bunimovich, L. A. [2010] "Which hole is leaking the most: A topological approach to study open systems," *Nonlinearity* **23**, 643–656.
2. Altmann, E. G. & Tél, T. [2009] "Poincaré recurrences and transient chaos in systems with leaks," *Phys. Rev. E* **79**, 016204.
3. Andreasen, J., Cao, H., Wiersig, J. & Motter, A. E. [2009] "Marginally Unstable Periodic Orbits in Semiclassical Mushroom Billiards," *Phys. Rev. Lett* **103**, 154101.
4. Armstead, D. N., Hunt, B. R. & Ott, E. [2004] "Power-law decay and self-similar distributions in stadium-type billiards," *Phys. D* **193**, 96–127.
5. Athreya, J. S. & Forni, G. [2008] "Deviation of ergodic averages for rational polygonal billiards," *Duke Math. J* **144**, 285–319.
6. Bálint, P. & Gouëzel, S. [2006] "Limit Theorems in the Stadium Billiard," *Commun. Math. Phys* **263**, 461–512.
7. Bálint, P. & Melbourne, I. [2008] "Decay of correlations and invariance principles for dispersing billiards with cusps, and related planar billiard flows," *J. Stat. Phys* **133**, 435–447.
8. Bálint, P. & Tóth, I. P. [2008] "Exponential decay of correlations in multi-dimensional dispersing billiards," *Ann. Henri Poincaré* **9**, 1309–1369.
9. Barreira, L. & Pesin Y. [2007] "Nonuniform hyperbolicity: Dynamics of systems with nonzero Lyapunov exponents," *Encycl. Math. Appl* **115**, Cambridge Univ. Press, Cambridge.
10. Bedaride, N. [2008] "Periodic billiard trajectories in polyhedra," *Forum. Geom* **8**, 107–120.
11. Broer, H. W. [2004] "KAM theory: the legacy of Kolmogorov's 1954 paper," *Bull. AMS (New Series)* **41** 507–521.
12. Bruin, H., Demers, M. & Melbourne, I. [2010] "Existence and convergence properties of physical measures for certain dynamical systems with holes," *Ergod. Th. Dyn. Sys* **30**, 687–728.
13. Bunimovich, L. A. [2008] "Relative volume of Kolmogorov-Arnold-Moser tori and uniform distribution, stickiness and nonstickiness in Hamiltonian systems," *Nonlinearity* **21**, T13–T17.
14. Bunimovich, L. A. & Del Magno, G. [2009] "Track billiards," *Commun. Math. Phys* **288**, 699–713.
15. Bunimovich, L. A. & Dettmann, C. P. [2005] "Open circular billiards and the Riemann hypothesis," *Phys. Rev. Lett* **94**, 100201.
16. Bunimovich, L. A. & Dettmann, C. P. [2007] "Peeping at chaos: Nondestructive monitoring of chaotic systems by measuring long-time escape rates," *EPL* **80**, 40001.
17. Bunimovich, L. A. & Grigo, A. [2010] "Focusing Components in Typical Chaotic Billiards Should be Absolutely Focusing," *Commun. Math. Phys* **293**, 127–143.
18. Bunimovich, L. A. & Yurchenko, A., "Where to place a hole to achieve the fastest escape rate," *Israel J. Math* (to appear).

19. Casati, G. & Prosen, T. [1999] "Mixing property of triangular billiards," *Phys. Rev. Lett* **83**, 4729–4732.
20. Chernov, N. I. & Markarian, R. [2006] "Chaotic billiards," *Mathematical surveys and Monographs* **127**, American Mathematical Society.
21. Chernov, N. I. & Simányi, N. [2010] "Upgrading the local ergodic theorem for planar semi-dispersing billiards," *J. Stat. Phys* **133**, 355–366.
22. Chernov, N. I. & Zhang, H. -K. [2009] "On statistical properties of hyperbolic systems with singularities," *J. Stat. Phys* **136**, 615–642.
23. Conrey, J. B. [2003] "The Riemann Hypothesis," *Not. Amer. Math. Soc* **50**, 341–353.
24. Conrey, J. B., Farmer D. W., Keating J. P., Rubinstein M. O. & Snaith, N. C. [2008] "Lower order terms in the full moment conjecture for the Riemann zeta function," *J. Number Theory* **128**,1516–1584.
25. Cristadoro, G. & Ketzmerick, R. [2008] "Universality of algebraic decays in hamiltonian systems," *Phys. Rev. Lett* **100** 184101.
26. Cserti, J., Hagymási, I. & Kormányos, A. [2009] "Graphene Andreev billiards," *Phys. Rev. B* **80**, 073404.
27. Demers, M. F., Wright, P. & Young, L. S. [2010] "Escape rates and physically relevant measures for billiards with small holes," *Commun. Math. Phys* **294**, 353–388.
28. Demers, M. F., & Young, L. -S. [2006] "Escape rates and conditionally invariant measures," *Nonlinearity* **19** 377–397.
29. Dettmann, C. P. [2000] "The Lorentz gas as a paradigm for nonequilibrium stationary states," in Hard ball systems and the Lorentz gas (Ed. D. Szasz) *Encyclopaedia of Mathematical Sciences* **101**, 315–365, Springer.
30. Dettmann, C. P. & Cohen, E. G. D. [2000] "Microscopic chaos and diffusion," *J. Stat. Phys* **101**, 775–817.
31. Dettmann, C. P. & Cohen, E. G. D. [2001] "Note on chaos and diffusion," *J. Stat. Phys* **103** 589–599.
32. Dettmann, C. P. & Georgiou, O. [2009] "Survival probability for the stadium billiard," *Phys. D* **238**, 2395–2403.
33. Dettmann, C. P., Morozov, G. V., Sieber, M. M. A. & Waalkens, H. [2009] "Unidirectional emission from circular dielectric microresonators with a point scatterer," *Phys. Rev. A* **80**, 063813.
34. Dolgopyat, D. & Fayad, B. [2009] "Unbounded orbits for semicircular outer billiard," *Ann. Henri Poincaré* **10**, 357–375.
35. Eckhardt, B., Russberg, G., Cvitanović, P., Rosenqvist, P. E., & Scherer, P. [1994] "Pinball scattering," in Casati, G. & Chirikov, B. V. (eds.) *Quantum chaos: Between order and disorder: A selection of papers*, Cambridge University Press, Cambridge.
36. Ezra, G. S., Waalkens, H. & Wiggins, S. [2009] "Microcanonical rates, gap times, and phase space dividing surfaces," *J. Chem. Phys* **130**, 164118.
37. Gaspard, P. [1998] *Chaos, scattering and statistical mechanics*, Cambridge University Press, Cambridge.
38. Gaspard, P. & Gilbert, T. [2008] "Heat Conduction and Fourier's Law by Consecutive Local Mixing and Thermalization," *Phys.Rev. Lett* **101**, 020601.

39. Grigo, A. [2009] *Billiards and statistical mechanics*, PhD thesis, Georgia Institute of Technology.
40. Gutkin, E. [1996] "Billiards in polygons: Survey of recent results," *J. Stat. Phys* **83**, 7–26.
41. Gutkin, E. [2003], "Billiard dynamics: A survey with the emphasis on open problems," *Reg. Chao. Dyn* **15**, 1–13.
42. Gutkin, E. [2010], "Geometry, topology and dynamics of geodesic flows on noncompact polygonal surfaces," *Reg. Chao. Dyn* **8**, 482–503.
43. Hassell, A., Hillairet, L. & Marzoula, J. [2009] "Eigenfunction concentration for polygonal billiards," *Commun. Part. Diff. Eq* **34**, 475–485.
44. Heller E. J. [2008] "Surface physics: Electron wrangling in quantum corrals," *Nature Physics* **4**, 443–444.
45. Jepps, O. G., Bianca, C. & Rondoni, L. [2008] "Onset of diffusive behavior in confined transport systems," *Chaos* **18**, 013127.
46. Kac, M. [1966] "Can one hear the shape of a drum?," *Amer. Math. Mon* **73**, 1–23.
47. Kaufmann, Z. & Lustfeld, H. [2001] "Comparison of averages of flows and maps," *Phys. Rev. E* **64**, 055206R.
48. Keating, J. P., Nonnenmacher, S., Novaes, M. & Sieber, M. [2008] "On the resonance eigenstates of an open quantum baker map," *Nonlinearity* **21**, 2591–2624.
49. Keller, G. & Liverani, C. [2009] "Rare events, escape rates and quasistationarity: Some exact formulae," *J. Stat. Phys* **135**, 519–534.
50. Lan, Y. [2010] "Cycle expansions: From maps to turbulence," *Commun. Nonlin. Sci. Numer. Simul* **15**, 502–526.
51. Lapidus, M. L. & Neimeyer, R. G. [2010] "Towards the Koch Snowflake fractal billiard: Computer experiments and mathematical conjectures," in More Tapas in Experimental Mathematics, (Eds. T. Amdeberham, L. A. Medina and V. H. Moll), American Mathematical Society.
52. Lazutkin, V. F. [1973] "Existence of caustics for the billiard problem in a convex domain," *Izv. Akad. Nauk SSSR Ser. Mat* **37**, 186–216.
53. Lopes, A. & Markarian, R. [1996] "Open billiards: cantor sets, invariant and conditionally invariant probabilities," *SIAM J. Appl. Math* **56**, 651–680.
54. Lu, W. T., Sridhar, S. & Zworski, M. [2003] "Fractal Weyl laws for chaotic open systems," *Phys. Rev. Lett* **91**, 154101.
55. Marklof, J. & Strömbergsson A. "The distribution of free path lengths in the periodic Lorentz gas and related lattice point problems," *Ann. Math* (to appear).
56. Marconi, U. M. B., Puglisi, A., Rondoni, L. & Vulpiani, A. [2008] "Fluctuation-dissipation: Response theory in statistical physics," *Phys. Rep* **461**, 111–195.
57. Nonnenmacher, S. [2008] "Some open questions in 'wave chaos'," *Nonlinearity*, **21**, T113–T121.
58. Novaes, M., Pedrosa, J. M., Wisniacki, D., Carlo, G. G. & Keating, J. P. [2009] "Quantum chaotic resonances from short periodic orbits," *Phys. Rev. E* **80**, 035202(R).

59. Petkov, V. & Stoyanov, L. [2009] "Correlations for pairs of periodic trajectories for open billiards," *Nonlinearity* **22**, 2657–2679.
60. Poli, C., Dietz, B., Legrand, O., Mortessagne, F. & Richter, A. [2009] "Avoided-level-crossing statistics in open chaotic billiards," *Phys. Rev. E* **80**, 035204R.
61. Pöllinger, M., O'Shea, D., Warken, F. & Rauschenbeutel, A. [2009] "Ultrahigh-Q tunable whispering-gallery-mode microresonator," *Phys. Rev. Lett* **103**, 053901.
62. Ramilowski, J. A., Prado, S. D., Borondo, F. & Farrelly, D. [2009] "Fractal Weyl law behavior in an open Hamiltonian system," *Phys. Rev. E* **80**, 055201(R).
63. Rapoport, A. & Rom-Kedar, V. [2008] "Chaotic scattering by steep repelling potentials," *Phys. Rev. E* **77**, 016207.
64. Schwartz, R. E. [2009], "Obtuse triangular billiards II: 100 Degrees Worth of Periodic Trajectories," *Exper. Math* **18**, 131–171.
65. Sinai, Y. G. [1970] "Dynamical systems with elastic reflections. Ergodic properties of dispersing billiards," *Russ. Math. Surv* **25**, 137–189.
66. Sinai, Y. G. [2010] "Chaos Theory Yesterday, Today and Tomorrow," *J. Stat. Phys* **138**, 2–7.
67. Sjöstrand, J. [1990] "Geometric bounds on the density of resonances for semiclassical problems," *Duke Math. J* **60**, 1–57.
68. Szász D. [2008] "Some challenges in the theory of (semi)-dispersing billiards," *Nonlinearity* **21**, T187–T193.
69. Tabachnikov, S. [2005] *Geometry and Billiards*, *Student Mathematical Library series* **30**, American Mathematical Society.
70. Troubeztkoy, S. [2010] "Typical recurrence for the Ehrenfest wind-tree model," arxiv:1004.3455.
71. Ulcigrai, C. [2009] "Absence of mixing in area-preserving flows on surfaces," arxiv:0901.4764.
72. Valdez, F. [2009] "Infinite genus surfaces and irrational polygonal billiards," *Geom. Dedicata* **143**, 143–154.
73. Waalkens, H., Burbanks, A. & Wiggins, S. [2005] "Escape from planetary neighbourhoods," *Mon. Not. R. Astron. Soc* **361**, 763–775.
74. Weaver, R. & Wright, M. (Eds.) [2010] "New Directions in Linear Acoustics and Vibration," Cambridge University Press.
75. Wiersig, J. & Main, J. [2008] "Fractal Weyl law for chaotic microcavities: Fresnel's laws imply multifractal scattering," *Phys. Rev. E* **77**, 036205.
76. Yang, H. L. & Radons, G. [2009] "Lyapunov modes in extended systems," *Phil. Trans. R. Soc. A***367** 3197–3212.

# Chapter 12

# Open Problems in the Dynamics of the Expression of Gene Interaction Networks

Larry S. Liebovitch[1] and Vincent Naudot[2]

[1] *Florida Atlantic University*
*Center for Complex Systems and Brain Sciences*
*Center for Molecular Biology and Biotechnology*
*Department of Psychology, Boca Raton, FL, 33431, USA*
*Email: liebovit@fau.edu*

[2] *Florida Atlantic University, Department of Mathematical Sciences*
*Boca Raton, FL, 33431, USA*
*Email: vnaudot@fau.edu*

Genes influence the expression of each other through a complex, nonlinear, dynamical network of interactions. There are a number of interesting open questions about what kind of information can be determined about the structure and dynamics of this network from limited experimental data.

**Keywords**: DNA, RNA, proteins, genes, networks, dynamics.

## Contents



## 12.1. Introduction

DNA (deoxyribonucleic acid) is the molecule that contains the programming code for building, maintaining, and altering living cells. The information in the genes encoded in DNA is transcribed into mRNA (messenger ribonucleic acid) which is then translated to synthesize the proteins that form physical structures and chemical reactions in the cell. But biology is much more complex and adaptive than just such a linear feed-forward chain. The DNA, mRNA, and proteins are all linked together in dynamical interacting feedback networks, see [Liebovitch *et al.*, 2010]. For example, some genes make proteins, called transcription factors, that bind onto DNA and increase or decrease the expression of genes. Some forms of RNA, called micro RNAs, target and destroy mRNA sequences, so that they are never translated into proteins. New technology can simultaneously measure the expression of tens of thousands of genes. Because of the comprehensive amount of data that can now be collected these methods are called "high throughput" technologies. They include the use of microarray chips that have small pieces of DNA that can bind RNA to identify which genes are being expressed, and real time polymerase chain reaction methods that amplify and then quantitatively assess which RNA is being expressed. Other new methods can also measure DNA-protein and protein-protein interactions, see [MacArthur *et al.*, 2009]. Therefore, in principle, it should be possible to construct a graph of all these interactions, where the nodes of the graphs are the genes and the edges between them are the net result of all the DNA, RNA, and protein interactions that link those genes together. Understanding the structure and dynamics of such a gene-gene interaction graph would give us great insight into how biology works and how best we could intervene in the network to restore the proper function compromised by disease, see [Liebovitch *et al.*, 2007]. However, the limited information provided by current experimental methods is not sufficient to determine the full structure and dynamics of this gene-gene interaction graph. The open problems presented below ask what is the information about the gene-gene interaction graph that can be determined from the experimental data of gene expression. To understand the problems described above our approach uses the notion of attractors in dynamical systems. Before stating the problems of this paper, we need to recall several concepts.

## 12.2. Attractors for Flows and Diffeomorphisms

When studying the evolution of a specific system from the theoretical point of view, we need to distinguish between continuous and discrete systems. The former is often the flow of vector field $\mathcal{X}$ defined on a manifold $\mathcal{M}$ and represented by an ordinary differential equation of the form

$$\dot{\mathbf{x}} = F(\mathbf{x})$$

where $\mathbf{x} \in \mathcal{M}$ and $F$ is a smooth function. The flow of $\mathcal{X}$ at the time $t \in \mathbb{R}$

$$\mathcal{X}_t : \mathcal{M} \to \mathcal{M}, \quad \mathbf{x} \mapsto \mathcal{X}_t(\mathbf{x})$$

consists in taking the value of the solution of the above equation with initial $\mathbf{x}$ at the time $t$.

It is sometimes not appropriate to model the evolution of a system using the approach from ODE and discrete systems are used to model the system. A discrete system is represented by a diffeomorphism

$$\Phi : \mathcal{M} \to \mathcal{M}, \ \mathbf{x} \mapsto \Phi(\mathbf{x})$$

that is a one to one differentiable map from which its Jacobian at each point is also invertible. It is well known that a flow is always a diffeomorphism. However, on the same manifold, a diffeomorphism is not always the time 1 of a vector field. For instance if one considers an orientation reversing diffeomorphism, such a situation cannot coincide with the time 1 of a vector field on the same manifold. Another difference observed in many models in biology, ecology and finance is when a vector field possesses a homoclinic orbit, that is an orbit which is bi-asymptotic to a singularity. In the case of a flow, if a point is homoclinic, then the whole orbit is also homoclinic. This means that we have a least a curve that is included in the intersection of the stable and unstable manifold. In the case of a diffeomorphism, this generically does not occur and we observe tangle between the stable and the unstable manifold.

For each system, the main interest is with stationary states, i.e., attractors. We say that a set $\mathcal{A}$ is an attractor for a vector field $\mathcal{X}$ if there is an open set $\mathcal{B} \supset \mathcal{A}$ called the basin of $\mathcal{A}$ such that for all $\mathbf{y} \in \mathcal{B}$

$$\lim_{t \to \infty} \operatorname{dist}\Big(\mathcal{X}_t(\mathbf{y}), \mathcal{A}\Big) = 0$$

where 'dist' is the Hausdorff distance. When considering a diffeomorphism, we say that $\mathcal{A}$ is a periodic attractor if there exists an integer $p \geq 1$ such that

$$\lim_{n \to \infty} \operatorname{dist}\Big(\Phi^{np}(\mathbf{y}), \mathcal{A}\Big) = 0$$

for all $\mathbf{y} \in \mathcal{B}$.

Finding explicit solutions for an ordinary differential equation is very hard in general. However, when concerning the same manifold, analyzing the dynamics of a flow is often easier than analyzing the dynamics of a diffeomorphism. In the next section we present a problem formulated by means of a linear diffeomorphism.

## 12.3. Statement of the Problem

Let $x$ be a vector whose elements are the expression levels of the set of genes. At each time step $n$ a function $A$ describes how these genes interact. Thus, $x_{n+1} = A(x_n)$. These time steps are repeated many times, so that from an initial state $x_0$, the final state

$$\lim_{n\to\infty} A^n(x_0) = x_f \tag{12.1}$$

Typically, only $x_f$ is known from the experimental data and both $x_0$ and $A$ are unknown. (Sometimes, there may be some experimental data known about a limited number of the time steps $x_n$ which provides a further extension of the simplest form of this problem.) There is not enough information to determine $A$ from $x_f$, see discussion below. But, that does not mean that we have no information about $A$. **The question is: What information can we determine about $A$ from just $x_f$?** These properties, for example, may include some of the group properties of the function $A$, or the statistical properties of $A$. The really hard part of this problem is to determine which are the properties of $A$ that can be determined from the corresponding properties of $x_f$. In what follows, we present a discussion to explain why this problem is hard to understand from the mathematical point of view.

### 12.3.1. *A First Attempt*

The initial value $x = x_0$ is unknown and so that we may say that it can be chosen randomly. This way we can analyze Eq. (12.1) for typical values of $x_0$ and ignore "exceptional cases". For simplicity, we shall assume that the field the matrix is written is the complex field $\mathbb{C}$. We then identify $A$ with its matrix written in a given basis $(e_1, \ldots, e_n)$ and write $A$ into its Jordan form:

$$A = M^{-1} \cdot D \cdot M$$

where $D = S[\mathrm{Id}+N]$, $S$ is a semi-simple matrix (i.e., diagonal in the present context), and $N$ is a nilpotent matrix, that is there exists $m \leq n$ such that

$N^m \equiv 0$. Moreover, Id is the identical matrix and $M$ is an invertible matrix. Solving Eq. (12.1) amounts to solving

$$\lim_{n\to\infty} M^{-1} \cdot S^n (\mathrm{Id} + N)^n \cdot M \cdot x_0 = x_f$$

Put

$$M \cdot x_0 = y_0, \quad y_f = M \cdot x_f.$$

The above equation takes the form

$$\lim_{n\to\infty} S^n (\mathrm{Id} + N)^n \cdot y_0 = y_f \tag{12.2}$$

Observe that

$$(\mathrm{Id} + N)^n = \mathrm{Id} + \tilde{N}_n$$

where $\tilde{N}_n$ is another Nilpotent matrix. From this consideration, it is clear that in the statement of Eq. (12.1) the vector $x_f$ is not arbitrary, but surely

$$x_f \in \mathrm{Im}(S)$$

that is the range of the matrix $S$. We introduce the set of eigenvalues of $S$, that is the element of $S$ on the diagonal (counted with their multiplicity)

$$\mathrm{Spect}(S) = \{\lambda_1, \ldots, \lambda_n\}$$

We can distinguish 4 difference cases.

[a] $|\lambda_j| < 1$: since $|\lambda_i|^n \to 0$ as $n \to \infty$, this implies that the corresponding entry for $y_f$ vanishes

[b] $|\lambda_j| > 1$ since $|\lambda_i|^n \to \infty$ as $n \to \infty$, this implies that the corresponding entry for $y_0$ vanishes. In the case $x_0$ (and therefore $y_0$) is chosen randomly, we must exclude this case. This situation is possible only in the case where we restrict the initial condition $x_0$ to a specific domain.

[c] $|\lambda_j| = e^{2i\pi\alpha}$ where $\alpha \in \mathbb{R}\backslash\mathbb{Q}$. In this case since the set

$$\{\lambda_j^m, \mid m \in \mathbb{N}\}$$

is dense in the unit circle, we do not have convergence of the sequences $(y_n)$ and $(x_n)$, therefore this case has to be rejected again.

[d] $|\lambda_j| = e^{2i\pi\alpha}$ where $\alpha \in \mathbb{Q}\backslash\mathbb{N}$. In this case we do not have convergence of the sequences $(y_n)$ and $(x_n)$. However, there exist an integer $q$ such that $\lambda_j^q = 1$ and as we will see in case [e], if we consider $A^q$ instead of $A$, we may have convergence of the sequences $(y_n)$ and $(x_n)$. In this case, $A^n$ will converge to a period attractor.

[e] $\lambda_j = 1$. In this case we claim that the restriction of $A$ on the corresponding eigenspace is the identity. If not, then the restriction of $A$ on that space takes the form $D(u) = u + N_j(u)$ where $u \in \mathbb{C}^k$, $k$ being the dimension of the eigenspace and $N_j$ is a nilpotent matrix. Therefore for all integer $n$ we have

$$D^n(u) = u + nN_j(u) + \mathcal{P}(N_j)(u)$$

where $\mathcal{P}(N)$ is a polynomial in the endomorphism $N$ i.e.,

$$\mathcal{P}(N_j) = \sum_{\ell=2}^{k-1} \beta_j N_j^\ell$$

This means that starting by a vector $u_0$ that we have no convergence of the sequence $(u_n)$ where $u_{n+1} = D(u_n)$ and since the components of $u_n$ are components of $y_n$ we then conclude that the sequences $(y_n)$ and $(x_n)$ do not converge unless $N \equiv 0$.

From the considerations above and after re-ordering the eigenspaces, we can conclude that the matrix $D$ takes the form

$$D = \begin{pmatrix} \mathrm{Id}_k & 0 \\ 0 & C \end{pmatrix}$$

where $\mathrm{Id}_k$ is the identical matrix on $\mathbb{C}^k$ and $C$ is a contraction matrix, i.e., there exists $0 < \alpha < 1$ such that

$$\|C(u)\| \leq \alpha \|u\|$$

We illustrate the above with the following examples.

### 12.3.2. *Examples*

**Example I.** Take the following $2 \times 2$ matrix

$$D = \begin{pmatrix} 1 & 0 \\ 0 & \alpha \end{pmatrix}$$

where $\alpha = 1$. It is clear that

$$\lim_{n\to\infty} D^n = \begin{pmatrix} 1 & 0 \\ 0 & 0 \end{pmatrix}$$

Take now

$$M = \begin{pmatrix} 3 & 7 \\ 2 & 5 \end{pmatrix}$$

A straightforward computation show that

$$A^n = M^{-1} \cdot D^n \cdot M = \begin{pmatrix} 15 - 14\alpha & 35 - 35\alpha \\ -6 + 6\alpha & -14 + 15\alpha \end{pmatrix} \to \begin{pmatrix} 15 & 35 \\ -6 & -14 \end{pmatrix}$$

as $n$ tends to $\infty$. It is clear from this example that by taking different initial vectors, we end up with different limits. For instance let us consider the following vectors:

$$e_1 = \begin{pmatrix} 1 \\ 0 \end{pmatrix}, \quad e_2 = \begin{pmatrix} 0 \\ 1 \end{pmatrix}$$

with is example we clearly verify that

$$A^n(e_1) \to \begin{pmatrix} 15 \\ -6 \end{pmatrix}$$

but

$$A^n(e_2) \to \begin{pmatrix} 35 \\ -4 \end{pmatrix}$$

Moreover, even if the limit exists, we cannot determine the value of $\alpha$. Finally the matrix $M$ is unknown. To make the problem clearer let us start with two additional specific examples.

**Example II.** Let $A$ be the $m$ x $m$ adjacency matrix of the gene-gene interaction graph for $m$ genes. Each element of $A_{ij}$, which is 0 or 1, determines if the gene $j$ influences the gene $i$. At each time step $x_{n+1} = Ax_n$ and thus the final expression state

$$x_f = \lim_{n\to\infty} A^n(x_0)$$

Given only $x_f$ we cannot determine the elements $A_{ij}$ of the matrix $A$. We cannot even determine all the eigenvalues or eigenvectors of $A$. However, perhaps surprisingly, we can determine some important properties of $A$. For example, $x_f$ is the eigenvector associated with the largest eigenvalue of $A$. Moreover, [Shehadeh *et al.*, 2006] showed through numerical simulations, for some specific cases, that there is a relationship between the statistics of the elements of $A$ and the statistics of $x_f$. Let the density of non-zero elements in $A$ be distributed uniformly in the rows of $A$, and the total number of non-zero elements in each row be proportional to

$$(\frac{1}{m-r})^{[1/(c-1)]}$$

where $r$ is the row number. Then, for the graph defined by $A$, the in-degree distribution $g(k)$, that is, the number of nodes receiving incoming edges from k nodes, has the power law form

$$g(k) \propto k^{-c}$$

This leads to a $x_f$ whose probability density function, $pdf(x)$, is a power law, namely,

$$pdf(x) \propto x^{-d}$$

where $d$ is a function of $c$. These results were compared to the experimental data of mRNA expression as measured by Affymetrix microarray chips. This was done by comparing the probability density function of the mRNA levels on the chips with those computed from the attractors of the matrices with different in-degree and out-degree distributions. Since these pdfs are distributions with long tails, we compared the slopes of those tails, on logarithmic-logarithmic plots, for both the experimental data and the computed attractors. Comparison with the experimental data suggests that $c \simeq d \simeq 2$, see [Shehadeh *et al.*, 2006] for more details. Thus, the statistical properties of the elements of $A$ define a type of graph with certain a degree distribution, which generates a certain statistical property in $x$, namely the $pdf(x)$. Again, the hard part of this problem is to define what properties of $A$ can be determined given only the resultant vector $x_f$. Even for this simplest case, where $A$ is an $m$ x $m$ matrix, this problem is beyond the scope of the usual matrix properties, such as eigenvectors or eigenvalues. There are also computational issues as the total number of interacting genes $m$ is of the order of 40,000. The problem is to define relevant properties (e.g. group, statistical, or other properties) that give us insight into the local or global properties of gene interactions as evidenced only by the relevant properties of the final state of the levels of gene expression. For example, these statistical properties might be the probability density distribution of the values of all the elements of the matrix, or of the elements in a row or column, or the probability density distribution of all the non-zero elements, or of the elements in a row or column, or the higher moments of such distributions in the whole matrix or in parts of the matrix.

**Example III.** If there are multiple experimental time series of the expression values as a function of time, then a set of basis functions can be used to identify dynamic functional linkages in the network defined by the adjacency matrix $A$. For example, singular value decomposition, SVD, has been used to identify limit cycles corresponding to harmonic oscillators from the mRNA expression data in the cell mitotic cycle in yeast, see [Alter, 2006]. SVD provides a set of orthogonal basis vectors and their eigenvalues that together span the details of the experimental data thereby summarizing the details of the data. They showed that some of these eigenvalues correlated with the activities of molecules previously identified to be the oscillators whose period corresponds to the changes in the cell, the mitotic cycle, as the cell makes copies of its internal constituents and then splits into two daughter cells. These results were shown to be robust to the perturbations and experimental errors in the experiments. The relationship of these basis functions to the other mathematical properties of $A$ and even more so, to the other mathematical dynamical properties on $A$, is not clear.

## 12.4. Experimental Information

For each of the specific problems described in the following section there are three different cases which correspond to different amounts or types of experimental data that may be available about $x_f$ (and also possibly $x_n$

**Case I.** There is only one experiment, starting from unknown initial levels of gene expression $x_0$, that converges to a known levels of gene expression in $x_f$.

**Case II.** There are multiple replications of the experiment (presumably each with different unknown initial conditions of the levels of gene expression $x_0$). Each experiment converges to the same levels of gene expression $x_f$, or the experiments converge to $p$ different final levels of gene expression $x_f^{\{1\}}, \ldots, x_f^{\{p\}}$. These $p$ attractors may represent some, or all, of the possible attractors of the system.

**Case III.** The levels of gene expression $x_n$ do not converge to a final steady state and information is available at multiple time points $x_n$ during the course of the experiments. These time points, may or may not, fully resolve the time behavior of the levels of gene expression. This case extends the simplest form of the problem to broader issues about the dynamics of this system.

## 12.5. Theoretical Models of Gene Interaction

For all of the three cases described above, the problem is to determine what properties of $A$ can be determined from what properties of $x_f$ (and also possibly $x_n$), when we make the following assumptions about the nature of $A$.

**Problem 12.1.** ***Determine A where A is the Adjacency Matrix.*** *This model, often commonly used in systems biology, assumes that genes influence each other through linear, additive interactions that do not depend on the state of other genes. That is, the contribution of gene $j$ to the expression of gene $i$ is proportional only to $x_j$ and the effect of all the other genes on gene $i$ is the sum of $A_{ij}x_j$.*

**Problem 12.2.** ***Determine A where A is a Nonlinear Function of the Gene Expression of Each Gene.*** *Biochemical reactions are typically not linear functions and it is more likely that the contribution of each gene $j$ to the expression of gene $i$ is a nonlinear function. A typical such function is that the rate of expression of $x_i$ is proportional to $x_j^h/[1+(x_j/x_{j0})^h]$, where $h$ is the Hill coefficient and $x_{j0}$ is a constant, if gene $j$ stimulates gene $i$, or $1/[1+(x_j/x_{j0})^h]$ if gene $j$ inhibits gene $i$. The Hill coefficient is a way of describing the nonlinearity in a chemical reaction. It describes how the amount of the product produced depends on the amount of the reactants in the chemical reaction. For typical biochemical reactions the Hill coefficient is in the range [1,4].*

**Problem 12.3.** ***Determine A where A is a Nonlinear Function of the Gene Expression of Multiple Genes.*** *Experimental evidence demonstrates that biology is often "context dependent", namely, that the effect produced by the binding of any one molecule A on molecule B is influenced by the presence of a third molecule C. For example, the effect of the binding of transcription factors on the regulatory region of a gene is influenced by the other transcription factors already bound at nearby sites on the DNA, see [Barash et al., 2003], In this case $A$ is a nonlinear function of two or more of the gene expression levels, $x_j$, $j = 1, 2, ...q$, $q \geq 2$.*

## 12.6. Conclusions

The levels of gene expression observed experimentally depend on interactions between many genes executed at the DNA, RNA, and protein levels.

There is insufficient information to determine the local and global properties of the gene-gene interaction network $A$ from the final expression $x_f$ of just a few experiments. But, that does not mean that no properties of $A$ can be determined. In fact, important mathematical properties of this network can be determined from that data. For example, [Shehadeh *et al.*, 2006] demonstrated that, for some networks, the statistical properties of the elements of the matrix form of $A$ are related to the statistical properties of $x_f$. The hard part is to determine, which existing or newly defined properties of these interaction functions $A$ can be determined from the existing or newly defined properties of $x_f$. If mathematical properties can be found that have important and useful biological relevance, they may give us deep insight into how these networks function and how we could alter them to cure diseases.

## References

1. Liebovitch, L. S., Shehadeh. L. A., Jirsa, V. K., Hütt M. T. & Marr, C. [2010] "Determining the Properties of Gene Regulatory Networks from Expression Data," *Handbook of Research on Computational Methodologies in Gene Regulatory Networks.* Edited by Das S, Caragea D, Welch S, Hsu WH, (IGI Global, Hershey PA), in press.
2. MacArthur, B. D., Ma'ayan, A. & Lemischka, R. [2009] "Systems biology of stem cell fate and cellular reprogramming," *Nature. Rev. Mol. Cell. Biol* **19**, 672–681.
3. Liebovitch, L. S., Tsinoremas, N.& Pandya, A. [2007] "Developing combinatorial multi-component therapies (CMCT) of drugs that are more specific and have fewer side effects than traditional one drug therapies," *Nonlinear. Biomedical. Phys* **1**–11, doi:10.1186/1753-4631-1-11.
4. Shehadeh, L. A., Liebovitch, L. S. & Jirsa, V. K. [2006] "Relationships between the global structure of genetic networks and mRNA levels measured by cDNA microarrays," *Phys. A* **364**, 297–314.
5. Alter, O. [2006] "Discovery of principles of nature from mathematical modeling of DNA microarray data," *Proc. Natl. Acad. (USA)* **103**, 16063–16064.
6. Barash, Y., Elidan, G., Friedman, N.& Kaplan, T. [2003] "Modeling dependencies in protein-DNA binding sites," *Proceedings of the Seventh Annual International Conference on Research in Computational Molecular Biology*Edited by Vingron M, Istrail S, Pevzner P, Waterman M, 28–37, Berlin, http://portal.acm.org/citation.cfm?id=640079

# Chapter 13

## How to Transform a Type of Chaos in Dynamical Systems?

Zeraoulia Elhadj[1] and J. C. Sprott[2]

[1]*Department of Mathematics, University of Tébessa, (12002), Algeria*
*E-mail: zeraoulia@mail.univ-tebessa.dz and zelhadj12@yahoo.fr*

[2]*Department of Physics, University of Wisconsin*
*Madison, WI 53706, USA*
*E-mail: sprott@physics.wisc.edu*

At the present time, strange attractors can be classified into three principal classes: hyperbolic, Lorenz-type, and quasi-attractors. The hyperbolic attractors are the limit sets for which Smale's Axiom A is satisfied and are structurally stable. Periodic and homoclinic orbits are dense and are of the same saddle type. The Lorenz-type attractors are not quite structurally stable, although their homoclinic and heteroclinic orbits are structurally stable (hyperbolic), and no stable periodic orbits appear under small parameter variations. The quasi-attractors are the limit sets enclosing periodic orbits of different topological types and structurally unstable orbits.

In this paper, some new open problems are proposed. These problems are related to transforming non-chaotic dynamical systems to a specific defined type of chaos (one of the three types defined above) and transforming a defined type of chaos to another defined one (also one of the three types defined above). These questions have not been previously addressed in the literature. Solving such a problem opens an interesting field in chaos studies concerned with the classification and determination of the type of chaos observed experimentally, proved analytically, or tested numerically in theory and practice.

**Keywords**: Chaotification, hyperbolic, Lorenz-type, and quasi-attractors, transforming dynamical systems.

### Contents

## 13.1. Introduction

In this paper, we state some new open problems related to the well known terminology *chaotification* or *anticontrol of chaos*, which is the reverse of suppressing chaos in a dynamical system. The aim of this process is to create or enhance the system complexity for some special novel, time- or energy-critical interdisciplinary application. Examples include high-performance circuits and devices, liquid mixing, chemical reactions, biological systems, crisis management, secure information processing, and critical decision-making in politics, economics, military applications, etc. Many chaotification methods have been proposed to create or enhance the system complexity for several novel, time- or energy-critical interdisciplinary applications as mentioned above. For example, chaotification schemes were presented for discrete mappings in [Li, 2004; Lai &Chen, 1997, 2003; Chen & Lai, 1997(a-b), 1998; Chen *et al.*, 1998; Lin *et al.*, 2002; Wang & Chen, 1999, 2000; Li *et al.*, 2002 (a-b-c); Zeraoulia & Sprott, 2010] using Lyapunov exponents, or by the use of several modified versions of the Marotto theorem [Marotto, 1978, 2005; Chen *et al.*, 1998; Lin *et al.*, 2002], or by the use of the Li-Yorke definition of chaos [Li & Yorke, 1975]. Also, many chaotification methods have been proposed to generate chaos in continuous-time systems including differential-geometry control [Chen, 2003; Lû *et al.*, 2002], time-delayed feedback [Wang & Chen, 2000], and switching piecewise-linear control [Lû *et al.*, 2002]. Recently, an effective strategy for anticontrolling chaos in continuous-time systems has been discussed [Yang *et al.*, 2005] using a new homogeneity-based approach with the $p$-normal forms of nonlinear systems.

## 13.2. Hyperbolification of Dynamical Systems

In this section, we give some new open problems concerning the transformation of a dynamical system, not necessarily hyperbolic, to a hyperbolic system that exhibits chaos. We call this procedure *hyperbolification*, by which we mean a method that creates system complexity by using a hyperbolicity criterion that is well developed in theory. Generally, the dynamics of a system is interesting if it has a closed, bounded, and hyperbolic attractor, i.e., the coexistence of highly complicated long-term behavior, sensitive dependence on initial conditions, and the overall stability of the orbit structure. Elsewhere, the dynamics is trivial from the viewpoint of chaos theory. For a better description of these problems, we give a short description of hyperbolicity theory with some major definitions and principal results. Indeed, the *Smale horseshoe map* [Smale, 1967] extracted from a study of relaxation oscillations by discerning a geometric picture in the horseshoe shape is the most important map in chaos theory characterized by the existence of infinitely many robust periodic orbits. Most known mathematical and physical models exhibit this phenomenon, i.e., these models are hyperbolic in the sense that the tangent space at each point splits into two complementary directions such that the derivative contracts one of these directions and expands the other at uniform rates. In other words, let $f : \Omega \subset \mathbb{R}^n \longrightarrow \mathbb{R}^n$ be a $C^r$ real function that defines a discrete map also called $f$ and $\Omega$ is a manifold. Hence (a) A point $x$ is a non-wandering point for the map $f$ if for every neighborhood $U$ of $x$ there is a $k \geq 1$ such that $f^k(U) \cap U$ is nonempty. (b) The set of all nonwandering points is called the *nonwandering set* of $f$. (c) An $f$-invariant subset $\Lambda$ of $\mathbb{R}^n$ satisfies $f(\Lambda) \subset \Lambda$. Then one has the following definitions given in [Abraham & Marsden, 1978]:

**Definition 13.1.** If $f$ is a diffeomorphism defined on some compact smooth manifold $\Omega \subset \mathbb{R}^n$, an $f$-invariant subset $\Lambda$ of $\mathbb{R}^n$ is said to be hyperbolic if there exists a $0 < \lambda < 1$ and a $c > 0$ such that

(1) $T_\Lambda \Omega = E^s \oplus E^u$, where $\oplus$ means the algebraic direct sum.
(2) $Df(x) E^s_x = E^s_{f(x)}$, and $Df(x) E^u_x = E^u_{f(x)}$ for each $x \in \Lambda$.
(3) $\|Df^k v\| \leq c\lambda^k \|v\|$ for each $v \in E^s$ and $k > 0$.
(4) $\|Df^{-k} v\| \leq c\lambda^k \|v\|$ for each $v \in E^u$ and $k > 0$.

where $E^s$, $E^u$ are, respectively, the stable and unstable submanifolds of the map $f$, i.e., the two $Df$-invariant submanifolds, and $E^s_x$, $E^u_x$ are the two $Df(x)$-invariant submanifolds. For a compact surface, a result of Plykin implies that there must be at least three holes for a hyperbolic attractor and it looks locally like a Cantor set (in the stable direction) $\times$ an interval (in the unstable direction). Also, hyperbolic attractors have dense periodic points and a point with dense orbits.

For continuous-time dynamical systems, a hyperbolic subset $\Lambda$ of $\mathbb{R}^n$ can be defined as follows: Let $f : \Omega \subset \mathbb{R}^n \longrightarrow \mathbb{R}^n$ be a $C^r$ real function that defines a continuous-time system also called $f$ and $\Omega$ is a manifold. Let $M$

be a closed 3-manifold and $f$ be a $C^1$ vector field with generating flow $f_t$, $t \in \mathbb{R}$. Given $p \in M$, then: (a) The $\omega$-limit set of $p$, denoted by $\omega(p)$ is the set of points $x \in M$ such that $x = \lim_{n\to\infty} f_{t_n}(p)$ for some sequence of real numbers $t_n \to \infty$. (b) A compact invariant set $\Lambda$ of $f$ is isolated if there exists an open set $\Lambda \subset U$ such that $\Lambda = \cap_{t\in\mathbb{R}} f_t(U)$. (c) If $f_t(U) \subset U$ for $t > 0$, then $\Lambda$ is an attracting set. (d) The topological basin of an attracting set $\Lambda$ is the set $W^s(\Lambda) = \{x \in M : \lim_{t\to+\infty} dist\,(f_t(x), \Lambda) = 0\}$. (e) A compact invariant set $\Lambda$ of $f$ is transitive if $\Lambda = \omega(p)$ for some $p \in \Lambda$. (f) A compact invariant set $\Lambda$ of $f$ is attracting if it realizes as $\cap_{t>0} f_t(U)$ for some compact neighborhood $U$ called the basin of attraction. (g) An attractor is a transitive attracting set. (h) An attractor, or repeller, is proper if it is not the whole manifold $M$. (l) An invariant set of $f$ is non-trivial if it is neither a periodic orbit nor a singularity.

Now assume that all basins of attraction are smooth manifolds with boundaries transverse to $f$. Then we have the following definition:

**Definition 13.2.** A compact invariant set $\Lambda$ of $f$ is hyperbolic if there exist positive constants $K$ and $\lambda$ and a continuous invariant tangent bundle decomposition $T_\Lambda M = E^s_\Lambda \oplus E^f_\Lambda \oplus E^u_\Lambda$, such that:

(i) $E^s_\Lambda$ is contracting, i.e., $\left\| Df_t(x) \left|_{E^s_x} \right. \right\| \leq K e^{-\lambda t}, \forall t > 0, \forall x \in \Lambda$.

(ii) $E^f_\Lambda$ is expanding, i.e., $\left\| Df_{-t}(x) \left|_{E^u_x} \right. \right\| \leq K e^{-\lambda t}, \forall t > 0, \forall x \in \Lambda$.

(iii) $E^f_\Lambda$ is tangent to $f$.

By this definition, it follows that (a) a hyperbolic attractor is an attractor that is simultaneously a hyperbolic set, (b) a hyperbolic repeller is a hyperbolic attractor for the time-reversed vector field, and (c) a closed orbit of $f$ is hyperbolic if it is hyperbolic as a compact invariant set of $f$. In this case, the nontrivial hyperbolic attractor $\Lambda$ is often called a *hyperbolic strange attractor*, and this is equivalent to the fact that $E^s_\Lambda \neq 0$ for the corresponding hyperbolic splitting of $\Lambda$. The above definition implies that in strange attractors of the hyperbolic type, all orbits in phase space are of the saddle type and the invariant sets of trajectories approach the original one in forward or backward time, i.e., the stable and unstable manifolds intersect transversally.

Now a hyperbolic set $\Lambda$ is locally maximal (or isolated) if there exists an open set $U$ such that $\Lambda$ is the union of all images of $U$ under the applications $f^n$, where $n \in \mathbb{Z}$. Note that the Smale horseshoe and the Plykin attractor are examples of locally maximal attractors (for the Smale horseshoe, every point for which all iterates lie in the rectangle $D$ define this map). In fact, if a $\Lambda$ is a hyperbolic set with a nonempty interior for a map $f$, then $f$ is Anosov if it is transitive, locally maximal, and $\Omega$ is a surface. These types of hyperbolic sets have properties including the following: shadowing, structural stability, Markov partitions, and SRB measures for attractors. When $\Lambda = \Omega$, then the diffeomorphism $f$ is called an *Anosov diffeomorphism* or *uniformly hyperbolic.* Thus we have the following definition:

**Definition 13.3.** (1) The map $f$ is uniformly hyperbolic or an *Axiom A* diffeomorphism if (a) The nonwandering set $\Omega(f)$ has a hyperbolic structure, (b) The set of periodic points of $f$ is dense in $\Omega(f)$, i.e., $Cl\,(Per\,(f)) = \Omega(f)$, the closure is the non-wandering set itself. (2) If $\varphi^t$ is a flow, then $\varphi^t : \Omega \subset \mathbb{R}^n \longrightarrow \mathbb{R}^n$ is an Anosov flow if for every $x \in \Omega$ there is a splitting of the tangent space $T_x\Omega = E^s \oplus E^0 \oplus E^u$, where $E^0$ is the flow direction and there are constants $C > 0$ and $\lambda \in (0,1)$ such that for every $t > 0$, one has $\|D\varphi^t(v)\| \leq c\lambda^t \|v\|$ for each $v \in E^s$, and $\|D\varphi^{-t}(v)\| \leq c\lambda^t \|v\|$ for each $v \in E^u$.

We notice that the stable and unstable subspaces $E^s$ and $E^u$ depend continuously on the point $x$ and they are invariant and interchanged when one passes from a map to its inverse. Furthermore, not every manifold admits an Anosov diffeomorphism or flow. For instance, the *hairy ball theorem* shows that there is no Anosov diffeomorphism on the 2-sphere. It is unknown whether the universal cover of a manifold that admits an Anosov diffeomorphism must be $\mathbb{R}^n$ for some $n$. Anosov diffeomorphisms are rather rare, and every known one of them is a generalization of automorphisms of a nilmanifold up to topological conjugacy such as the Smale horseshoe. An example of Anosov flows is the *free-particle motion*, or the mechanical system called *geodesic flow* on a compact surface (it looks locally like a mountain-pass landscape or the inner rim of a doughnut) of negative curvature. The main reason is that at every point the phase space can be decomposed into contracting and expanding directions. The most important features resulting from hyperbolicity are the coexistence of highly complicated long-term behavior, sensitive dependence on initial conditions, and the overall stability of the orbit structure. In fact, Axiom A diffeomorphisms serve as a model for the general behavior at a transverse homoclinic point, where the stable and unstable manifolds of a periodic point intersect, i.e., the study of homoclinic bifurcations [Anosov, 1967]. More generally, if $M$ is a $C^\infty$ compact manifold without a boundary and if $\mathcal{F}^1(M)$ denotes the set of diffeomorphisms $f \in Diff^1(M)$ having a $C^1$ neighborhood $\mathcal{U}$ such that all the periodic points of every $g \in \mathcal{U}$ are hyperbolic, then a proof of a conjecture of Mãné given in [Mañé, 1982] claiming that every element of $\mathcal{F}^1(M)$ satisfies Axiom A was given in [Hayashi, 1992].

To define the *Morse-Smale diffeomorphisms*, we need the definition of the *Kupka-Smale diffeomorphisms* given by the following:

**Definition 13.4.** The diffeomorphism $f$ is Kupka-Smale if all its periodic orbits are hyperbolic and moreover if for any periodic points $p$ and $q$, the unstable manifold $W^u(p)$ of $p$ and the stable manifold $W^s(q)$ of $q$ are in a general position, i.e., at any intersection point $x \in W^u(p) \cap W^s(q)$, we have $T_xM = T_xW^u(p) + T_xW^s(q)$.

It was shown in [Kupka, 1963; Smale, 1963] that for any $r \geq 1$, the set of Kupka-Smale $G_{KS}$ diffeomorphisms is a dense subset of diffeomorphisms defined in $\Omega$ and of class $C^r$ denoted by $Diff^r(\Omega)$ and equipped by the $C^r$-topology. As recent result, it was proved in [Pujals, 2008] for a given topologically hyperbolic attracting set of a smooth three-dimensional Kupka-Smale diffeomorphism, that either the set is hyperbolic or the diffeomorphism is $C^1$-approximated by another one exhibiting either a heterodimensional cycle or a homoclinic tangency. Thus the Morse-Smale diffeomorphism is defined by the following:

**Definition 13.5.** The diffeomorphism $f$ is Morse-Smale if it is a Kupka-Smale diffeomorphism whose nonwandering set $\Omega(f)$ (or equivalently whose chain-recurrent set $R(f)$) is finite.

If $C^r(M)$ is the space of $C^r$ vector fields, for any $r \geq 1$, and $C^r_\mu(M)$ is the subset of divergence-free vector fields defining incompressible (or conservative) flows, then the following result was proved in [Araujo & Bessa, 2008]:

**Theorem 13.1.** *(a) There exists a generic subset $\mathcal{R} \subset C^1_\mu(M)$ such that for $f \in \mathcal{R}$,*

- *either $f$ is Anosov,*
- *or else for Lebesgue almost every $p \in M$ all the Lyapunov exponents of $f^t$ are zero.*

*(b) Let $\varepsilon > 0$ be an open subset $U$ of $M$ and a non-Anosov vector field $f \in C^1_\mu(M)$ be given. Then there exists a $g \in C^1_\mu(M)$ such that $g$ is $C^1$-$\varepsilon$-close to $g$ and $g^t$ has an elliptic closed orbit intersecting $U$.*

*(c) There exists an open and dense subset $G \subset C^2_\mu(M)$ such that for every $f \in G$ with a regular invariant set $\Lambda$ (not necessarily closed) satisfying:*

- *the linear Poincaré flow over $\Lambda$ has a dominated decomposition; and*

• *$\Lambda$ has positive volume: $\mu(\Lambda) > 0$; then $f$ is Anosov, and the closure of $\Lambda$ is the whole of $M$.*

The most important property of a hyperbolic diffeomorphism is the structure of its invariants, in particular, corresponding iterated attractors, especially hyperbolic attracting sets with an isolating neighborhood or Lyapunov-stable attractors, which implies that all unstable manifolds lie in the attractor. Examples of a such situation are the Plykin attractor and the Smale-Williams solenoid.

**Definition 13.6.** (a) A hyperbolic set is defined to be a compact invariant set $\Lambda$ of a diffeomorphism $f$ such that the tangent space at every $x \in \Lambda$ admits an invariant splitting that satisfies the contraction and expansion conditions described in (3) and (4) of Definition 1.

(b) A locally maximal hyperbolic set is the largest invariant set in a small neighborhood of the hyperbolic set.

Hyperbolic flows can be obtained from diffeomorphisms using the so-called *special flows* constructed as follows: (a) Choose a hyperbolic diffeomorphism $f$ on a space $\Omega$. (b) Define using the unit-speed upward motion on $\{(x,t) : x \in \Omega, 0 \leq t \leq \varphi(x)\}$, the special flow over $f$ and under a *roof function* $\varphi$. The *suspension flow* is a special flow under the function $\varphi = 1$. This flow is never topologically mixing because at integer times the image of $\Omega \times \left(0, \frac{1}{2}\right)$ is disjoint from $\Omega \times \left(\frac{1}{2}, 1\right)$. But, for a generic roof function, the corresponding special flow over a topologically mixing Anosov flow is itself topologically mixing.

In what follows, we give some examples of hyperbolic systems. The first example is the Anosov diffeomorphisms on the torus $\mathbb{T}^n$. This includes Anosov automorphisms (*toral automorphisms*) defined as follows:

**Definition 13.7.** An automorphism is a hyperbolic, linear diffeomorphism on the torus $\mathbb{T}^n = \mathbb{R}^n/\mathbb{Z}^n$ into itself.

Toral automorphisms have the following properties (and for any transitive Anosov diffeomorphisms) that can be obtained using symbolic model as shown in [Yoccoz, 1993]: (a) The periodic points are dense. (b) There exists a point whose orbit is dense. (c) There are many ergodic invariant probability measures with full support. For example, the map defined by the matrix $\begin{pmatrix} 1 & 1 \\ 1 & 0 \end{pmatrix}$ and its square $\begin{pmatrix} 2 & 1 \\ 1 & 1 \end{pmatrix}$ has an entropy of $\log \frac{3+\sqrt{5}}{2}$. Due to the result proved in [Manning, 1974], it is sufficient to consider only the automorphisms on the torus $\mathbb{T}^n$ instead of nonlinear Anosov mappings since every Anosov diffeomorphism $f$ of $\mathbb{T}^n$ such that $\Omega(f) = \mathbb{T}^n$ is topologically conjugate to some Anosov automorphism of $\mathbb{T}^n$. It was proved in [Arrowsmith, 1990] that Anosov diffeomorphisms have complicated non-wandering sets. That is, a point $\theta \in \mathbb{T}^n$ is a periodic point of the Anosov automorphism $f : \mathbb{T}^n \to \mathbb{T}^n$ if and only if $\theta = \pi(x)$ where $x \in \mathbb{R}^n$ has rational coordinates. The map $\pi$ is defined in some way and is related to the automorphism $f : \mathbb{T}^n \to \mathbb{T}^n$. In this case, the map $f$ has infinitely many periodic points that are dense in the torus $\mathbb{T}^n$. All these points lie in the non-wandering set $\Omega(f)$ of $f$, and, since $\Omega(f)$ is closed, and one concludes that $\Omega(f) = \mathbb{T}^n$. Furthermore, it was shown in [Mather, 1968] that Anosov diffeomorphisms on $\mathbb{T}^n$, $n \geq 2$, are structurally stable on a compact manifold whose non-wandering set contains infinitely many points, i.e., Anosov diffeomorphisms on $\mathbb{T}^n$ are structurally stable in $Diff^1(\mathbb{T}^n)$, which implies that the dynamics on $\Omega(f)$ are very complicated, involving infinitely many hyperbolic periodic orbits densely distributed over the torus $\mathbb{T}^n$. A recent result on the topic of hyperbolic automorphisms of the 2-torus is given in [Anosov *et al.*, 2008] where the existence of an isomorphism between a deterministic dynamical system and a random process was established via an example of the circle expanding map (a map is called expanding if the length of any tangent vector field grows exponentially under the action of the differential of the map). A classification of hyperbolic toric automorphisms was also done, and the notion of the simplest Markov partitions was discussed in a new way with their suggested classification.

The second example is the so-called *Blaschke product* defined by

$$B(z) = \theta_0 \left(\frac{z - a_1}{1 - z\bar{a}_1}\right)\left(\frac{z - a_2}{1 - z\bar{a}_2}\right) \cdots \left(\frac{z - a_n}{1 - z\bar{a}_n}\right) \tag{13.1}$$

where $n \geq 2$, $a_i \in \mathbb{C}$; $|a_i| < 1, i = 1, \ldots, n$ and $\theta_0 \in \mathbb{C}$ with $|\theta_0| = 1$. The map (13.1) is rational in $\mathbb{C}$, and it is an analytic function in a neighborhood of the unit disc $\mathbb{D}$. The map $B$ maps the unit circle $\mathbb{T}$ to itself. In [Pujals *et al.*, 2006], the family of Blaschke products given by $\{B_\theta\}_{\theta \in \mathbb{T}} = \{\theta B\}_{\theta \in \mathbb{T}}$ was considered, and it was proved that this family is expanding or has a unique attracting or indifferent fixed point.

The third example is the *Bernoulli map* defined by

$$\phi_{n+1} = 2\phi_n(\bmod 2\pi) \tag{13.2}$$

The map (13.2) is expanding, exhibits *homogeneous chaotic dynamics* as shown in [Sinai, 1979; Devaney, 1989; Ott, 1993; Katok & Hasselblatt,

1995], and it transforms the angle variable $\phi$ is a non-uniform way, but it must be monotonous and possess the characteristic topological property. The fourth example is the *Arnold cat map* [Anosov, 1967] given by

$$\begin{cases} x_{n+1} = x_n + y_n, \ (\text{mod}1) \\ y_{n+1} = x_n + 2y_n, \ (\text{mod}1) \end{cases} \tag{13.3}$$

which is the Anosov torus $\mathbb{T}^2$. The map (13.3) is conservative, i.e., the *cat face* conserves under iteration the area of any domain in the $(x, y)$ plane. The map (13.3) has a hyperbolic chaotic attractor [Arnold, 1988; Bunimovich, 2000] since it has two Lyapunov exponents expressed via eigenvalues of the matrix associated with it, i.e., $\Lambda_1 = \frac{\ln(3+\sqrt{5})}{2} = 0.9624$ and $\Lambda_2 = -\frac{\ln(3+\sqrt{5})}{2} = -0.9624$, and their sum vanish. Finally, the second iteration of the Fibonacci map yields the Arnold cat map.

In the following, we notice that the hyperbolic theory of dynamical systems is widely used for characterizing the chaotic behavior of realistic nonlinear systems, but it has never been applied to any physical object with a continuous-time dynamical system. Generally, most known physical systems do not belong to the class of systems with hyperbolic attractors. Since hyperbolic strange attractors are robust (structurally stable) [Mira, 1997], it is very interesting to find physical examples of hyperbolic chaos, i.e., noise generators and transmitters in chaos-based communications. Recently, some continuous-time dynamical systems were constructed and confirmed to be hyperbolic. The proof was given on the basis of the corresponding Poincaré mapping [Kuznetsov & Seleznev, 2006]. The method most often used for such a construction is based on the use of coupled self-sustained oscillators with alternating excitation and numerical calculation of diagrams illustrating the phase transfer [Hunt, 2000; Belykh *et al.*, 2005; Kuznetsov, 2005; Kuznetsov & Seleznev, 2006; Isaeva *et al.*, 2006; Kuznetsov & Pikovsky, 2007; Kuznetsov, 2008; Kuznetsov & Pikovsky, 2008; Kuznetsov & Ponomarenko, 2008] where additional coupling permits the transfer of phases simultaneously from one partner to the other in order to obtain a desired chaotic (hyperbolic) map on a circle or on a torus. Notice that some of the constructed continuous-time hyperbolic systems have four variables such as the system in [Kuznetsov & Seleznev, 2006] given by

$$\begin{cases} x' = -2\pi u + (h_1 + A_1 \cos 2\pi t/N)\, x - \frac{1}{3}x^3 \\ u' = 2\pi\, (x + \varepsilon_2 y \cos 2\pi t) \\ y' = -4\pi v + (h_2 - A_2 \cos 2\pi t/N)\, y - \frac{1}{3}y^3 \\ v' = 4\pi \left(y + \varepsilon_1 x^2\right) \end{cases} \tag{13.4}$$

and with six variables such as the system in [Kuznetsov & Pikovsky, 2007] given by:

$$\begin{cases} x_1' = \omega_0 y_1 + \left(1 - a_1 - \frac{1}{2}a_2 - 2a_3\right) x_1 + \varepsilon \left(x_2 y_3 + x_3 y_2\right) \\ y_1' = -\omega_0 x_1 + \left(1 - a_1 - \frac{1}{2}a_2 - 2a_3\right) y_1 \\ x_2' = \omega_0 y_2 + \left(1 - a_2 - \frac{1}{2}a_3 - 2a_1\right) x_2 + \varepsilon \left(x_1 y_3 + x_3 y_1\right) \\ y_2' = -\omega_0 x_2 + \left(1 - a_2 - \frac{1}{2}a_3 - 2a_1\right) y_2 \\ x_3' = \omega_0 y_3 + \left(1 - a_3 - \frac{1}{2}a_1 - 2a_2\right) x_3 + \varepsilon \left(x_2 y_1 + x_1 y_2\right) \\ y_3' = -\omega_0 x_3 + \left(1 - a_3 - \frac{1}{2}a_1 - 2a_2\right) y_3, \end{cases} \tag{13.5}$$

and with eight variables such as the system in [Isaeva *et al.*, 2006] given by

$$\begin{cases} x'' - \left[A \cos \frac{2\pi t}{T} - x^2\right] x' + \omega_0^2 x = \varepsilon z \cos \omega_0 t \\ y'' - \left[A \cos \frac{2\pi t}{T} - y^2\right] y' + \omega_0^2 y = \varepsilon w \\ z'' - \left[-A \cos \frac{2\pi t}{T} - z^2\right] z' + 4\omega_0^2 z = \varepsilon x y \\ w'' - \left[-A \cos \frac{2\pi t}{T} - w^2\right] w' + \omega_0^2 w = \varepsilon x \end{cases} \tag{13.6}$$

Finally, from the above description, we can conclude that hyperbolic attractors are the limit sets for which Smale's *Axiom A* is satisfied and they are structurally stable and periodic with homoclinic orbits that are dense and of the same saddle type, i.e., the same index (the same dimension for their stable and unstable manifolds) [Ott, 1993; Katok & Hasselblatt, 1995]. In fact, hyperbolic chaos is often called *true chaos* from the rigorous mathematical point of view and is characterized by a homogeneous and topologically stable structure as shown in [Anosov, 1967; Smale, 1967; Ruelle & Takens, 1971; Guckenheimer & Holmes, 1983].

**Problem 13.1.** *Let $f : \Omega \subset \mathbb{R}^n \longrightarrow \mathbb{R}^n$ be a $C^r$ real function that defines a discrete map also called $f$, not necessarily hyperbolic, and $\Omega$ is a manifold. Find a controller $u$ such that the function $f + u : \Omega' \subset \mathbb{R}^n \longrightarrow \mathbb{R}^n$ defines a non-trivial hyperbolic (true) chaotic map in the sense of Definition 1. Or find another method to do that. Here $\Omega'$ is the domain of definition of $f+u$.*

**Problem 13.2.** *Let $f : \Omega \subset \mathbb{R}^n \longrightarrow \mathbb{R}^n$ be a $C^r$ real function that defines a continuous-time system also called $f$, not necessarily hyperbolic, and $\Omega$ is a manifold. Find a controller $u$ such that the function $f+u : \Omega' \subset \mathbb{R}^n \longrightarrow \mathbb{R}^n$ defines a non-trivial hyperbolic (true) chaotic continuous-time system in the sense of Definition 2. Or find another method to do that. Here $\Omega'$ is the domain of definition of $f + u$.*

## 13.3. Transforming Dynamical Systems to Lorenz-Type Chaos

In this section, we give some new open problems concerning the transformation of a dynamical system, not necessarily of the Lorenz-type, to a Lorenz-type system that exhibits chaos. Before doing that, we will give a short description with some definitions and results concerning the theory of Lorenz-type systems. Indeed, the Lorenz-type attractors (close in their structure and properties to robust hyperbolic attractors) are not quite structurally stable, although their homoclinic and heteroclinic orbits are structurally stable (hyperbolic), and no stable periodic orbits appear under small parameter variations. Examples of such a case are the Lorenz system itself [Lorenz, 1963], the Belykh attractor [Belykh, 1982-1995], and the Lozi attractor [Lozi, 1978], considered as a Lorenz-type attractor realized in a two-dimensional map (at least one of the three conditions of hyperbolicity given in Definition 1 is violated). These attractors are considered as examples of *truly* strange attractors [Shil'nikov, 1981], i.e., for example in the Lorenz attractor, all trajectories are saddles, and the variation of parameters does not create stable points or cycles [Bykov & Shil'nikov, 1989; Afraimovich *et al.*, 1982]. This implies that for certain sets of parameter values, the Lorenz system has the key properties of hyperbolic attractors, which confirms that this system is not completely hyperbolic. For this reason, the Lorenz system is sometimes called *quasi-hyperbolic* [Afraimovich *et al.*, 1977; Mischaikow & Mrozek, 1995-1998]. The Lorenz-type systems differ from robust hyperbolic attractors by the existence of a local violation of homogeneity (the birth of non–robust homoclinic trajectories) due to the presence of singular phase trajectories belonging to another saddle type, i.e., saddle equilibrium states that have different dimensions of its manifolds, or separatrix circuits, just like the Lorenz attractor, which includes a denumerable set of separatrix loops of the saddle equilibrium state as shown in [Williams, 1977; Afraimovich *et al.*, 1977; Shil'nikov, 1980]. From the viewpoint of bifurcations, an attractor is a Lorenz-type if it has non-dangerous trajectories with a zero measure on the attractor, i.e., their appearance and disappearance should not lead to the birth of stable trajectories and affect the structure of the chaotic hyperbolic set. In fact, Lorenz-like attractors can appear from singular cycles, and they are characterized in a robust way by the presence of infinitely many periodic orbits in any neighborhood of a singularity. Hence, from the above considerations, the Lorenz-like attractors can be defined as follows [Araujo & Pacifico, 2008]:

**Definition 13.8.** A Lorenz-like attractor is an attractor with the following characteristics:

(1) It is a robust, transitive attractor, which is not hyperbolic.

(2) The origin $(0, 0, 0)$ accumulates hyperbolic periodic orbits.

(3) The attractor has sensitive dependence on initial conditions (is chaotic).

Recent results can be found in [Komuro, 1984; Morales *et al.*, 1998; Anishchenko *et al.*, 2002; Klinshpont *et al.*, 2005; Bautista & Morales, 2006; Klinshpont, 2006; Alves *et al.*, 2007; Araujo & Pacifico, 2007; Araujo & Bessa, 2008; Araujo & Pacifico, 2008]. Now let $\Lambda$ be an attractor of a Lorenz-like system in a compact boundaryless 3-manifold $M$. Let $C^1(M)$ denote the set of $C^1$ vector fields on $M$ endowed with the $C^1$ topology. Thus it was claimed in [Araujo & Pacifico, 2008] that Lorenz-like attractors have the following properties: (1) There is an invariant foliation whose leaves are forward contracted by the flow. (2) There is a positive Lyapunov exponent for every orbit. (3) They are expansive and thus sensitive to initial data. (4) They have zero volume if the flow is $C^2$. (5) There is a unique physical measure whose support is the whole attractor and which is the equilibrium state with respect to the center-unstable Jacobian.

The best known example of a Lorenz-type system is the original Lorenz system itself [Lorenz, 1963] given by

$$\begin{cases} x' = \sigma\,(y - x) \\ y' = rx - y - xz \\ z' = -bz + xy \end{cases} \tag{13.7}$$

which is the first known case of a fully non-hyperbolic persistent and transitive attractor for all sufficiently small variations of the parameters $\sigma = 10, b = \frac{8}{3}$, and $r = 28$ in which periodic and homoclinic orbits are everywhere dense.

The second example of this situation is Lozi map [Lozi, 1978] given by

$$\begin{cases} x_{n+1} = 1 - a\,|x_n| + y_n \\ y_{n+1} = bx_n \end{cases} \tag{13.8}$$

where $a$ and $b$ are real parameters. The Lozi map (13.8) is considered as a Lorenz-type attractor realized in a two-dimensional map. It is known that *dangerous* tangencies are an essential characterization of the dynamics of the Hénon map (13.9) [Hénon, 1976] below. This phenomenon can be avoided by using piecewise-smooth functions [Banerjee & Grebogi, 1999] as in the case of the Lozi map (13.8). We notice that the Lozi strange attractor has a nearly hyperbolic structure (it is a piecewise affine uniformly hyperbolic map) except on the $y$-axis where the Lozi map (13.8) is not differentiable, and by the *influence of the singularities* on the $y$-axis, the dynamics of the Lozi map (13.8) is quite delicate [Misiurewicz, 1980]. In fact, the Lozi map (13.8) is close to the Lorenz attractor given by (13.7). A detailed study of the Lozi mapping (almost all known results in the literature) can be found in the forthcoming book [Zeraoulia, 2011].

Now we state two open problems on the topic of transforming an arbitrary dynamical system to a Lorenz-type system:

**Problem 13.3.** *Let $f : \Omega \subset \mathbb{R}^n \longrightarrow \mathbb{R}^n$ be a $C^r$ real function that defines a discrete map also called $f$, not necessarily of the Lorenz-type, and $\Omega$ is a manifold. Find a controller $u$ such that the function $f+u : \Omega' \subset \mathbb{R}^n \longrightarrow \mathbb{R}^n$ defines a non-trivial Lorenz-type map in the sense of Definition 11, or find another method to do that. Here $\Omega'$ is the domain of definition of $f+u$.*

**Problem 13.4.** *Let $f : \Omega \subset \mathbb{R}^n \longrightarrow \mathbb{R}^n$ be a $C^r$ real function that defines a continuous-time system also called $f$, not necessarily of the Lorenz-type, and $\Omega$ is a manifold. Find a controller $u$ such that the function $f+u : \Omega' \subset \mathbb{R}^n \longrightarrow \mathbb{R}^n$ defines a non-trivial Lorenz-type continuous-time system in the sense of Definition 11, or find another method to do that. Here $\Omega'$ is the domain of definition of $f+u$.*

**Problem 13.5.** *In particular, let $f : \Omega \subset \mathbb{R}^n \longrightarrow \mathbb{R}^n$ be a $C^r$ real function that defines a Lorenz-type discrete map also called $f$, and $\Omega$ is a manifold. Find a controller $u$ such that the function $f+u : \Omega' \subset \mathbb{R}^n \longrightarrow \mathbb{R}^n$ defines a non-trivial hyperbolic map in the sense of Definition 1, or find another method to transform a Lorenz-type chaotic map to a hyperbolic chaotic map, and vice versa. Here $\Omega'$ is the domain of definition of $f+u$.*

**Problem 13.6.** *In particular, let $f : \Omega \subset \mathbb{R}^n \longrightarrow \mathbb{R}^n$ be a $C^r$ real function that defines a Lorenz-type continuous-time system also called $f$, and $\Omega$ is a manifold. Find a controller $u$ such that the function $f+u : \Omega' \subset \mathbb{R}^n \longrightarrow \mathbb{R}^n$ defines a non-trivial hyperbolic continuous-time system in the sense of Definition 2, or find another method to transform a Lorenz-type chaotic continuous-time system to a hyperbolic chaotic continuous-time system, and vice versa. Here $\Omega'$ is the domain of definition of $f+u$.*

## 13.4. Transforming Dynamical Systems to Quasi-Attractor Systems

In this section, we give some new open problems concerning the transformation of a dynamical system, not necessarily a quasi-attractor, to a quasi-attractor system that exhibit chaos. Before doing that, we will give a short description with some definitions and results concerning the theory of quasi-attractor systems. Indeed, *quasi-attractors* are the limit sets enclosing periodic orbits of different topological types (for example stable and saddle periodic orbits) and structurally unstable orbits. Most known observed chaotic attractors are quasi-attractors [Afraimovich & Shil'nikov, 1983; Lichtenberg & Lieberman, 1983; Rabinovich, 1984; Schuster, 1984; Neimark & Landa, 1989; Anishchenko, 1990; Anishchenko, 1995]. This type of attractor has applications such as in neurodynamics [Fujii *et al.*, 2007(a-b)]. These attractors have the following important properties: (1) They have separatrix loops of saddle-focuses or homoclinic orbits of saddle cycles at the point of tangency of their stable and unstable manifolds because they enclose non-robust singular trajectories that are *dangerous*. (2) A map of *Smale's horseshoe-type* appears in the neighborhood of their trajectories. This map contains both a non-trivial hyperbolic subset of trajectories and a denumerable subset of stable periodic orbits (Shil'nikov's theorems [Shil'nikov, 1965; Gavrilov & Shil'nikov, 1973] and Newhouse's theorem [Newhouse, 1980]). (3) The quasi-attractor is the unified limit set of the whole attracting set of trajectories including a subset of both chaotic and stable periodic trajectories, which have long periods and weak basins of attraction and stability regions. This is a result of the effect that this attractor is *holed* by a set of basins of attraction of different periodic orbits. (4) The basins of attraction of stable cycles are very narrow. (5) Some orbits do not ordinarily reveal themselves in numerical simulations except in some large stability windows where they are clearly visible.

Hence, quasi-attractors have a very complex structure of embedded basins of attraction in terms of initial conditions and a set of bifurcation parameters of non-zero measure. The principal cause of this complexity is the homoclinic tangency of stable and unstable manifolds of saddle points in the Poincaré section [Gavrilov & Shil'nikov, 1973; Afraimovich *et al.*, 1983; Afraimovich & Shil'nikov, 1983]. This phenomenon implies that a rigorous mathematical description of quasi-attractors is still an open problem because almost all nonhyperbolic attractors are obscured by noise. Some new numerical and analytical results on this topic can be found in [Luchinski,

1999; Changming, 2005; Wang & Fang, 2006]. The necessary and sufficient condition for the existence of a quasi-attractor was given in [Zuo & Wang, 2007] by generalizing some results of Conley based on the connection among the attractor, the attractor neighborhood, and the domain of influence. Some examples of quasi-attractors include the Hénon map [Hénon, 1976] given by

$$\begin{cases} x_{n+1} = 1 - ax_n^2 + y_n \\ bx_n \end{cases} \tag{13.9}$$

the Anishchenko-Astakhov oscillator [Anishchenko, 1990, 1995] given by

$$\begin{cases} x' = mx + y - xz \\ y' = -x \\ z' = -g.z + g.\begin{cases} 1, x > 0 \\ 0, x \leq 0 \end{cases} .x^2 \end{cases} \tag{13.10}$$

the Strelkova-Anishchenko map [Strelkova & Anishchenko, 1997] (two coupled logistic maps) given by

$$\begin{cases} x_{n+1} = 1 - \alpha x_n^2 + \gamma\left(y_n - x_n\right) \\ y_{n+1} = 1 - \alpha y_n^2 + \gamma\left(x_n - y_n\right) \end{cases} \tag{13.11}$$

and Chua's circuit [Matsumoto *et al.*, 1985] given by

$$\begin{cases} x' = \alpha\left(y - \left(m_1 x + \frac{1}{2}(m_0 - m_1)\left(|x+1| - |x-1|\right)\right)\right) \\ y' = x - y + z \\ z' = -\beta y \end{cases} \tag{13.12}$$

A detailed study of the Hénon mappings and the Chua's circuit (almost all known results in the literature) can be found in the book [Zeraoulia & Sprott, 2010].

Now, we state six open problems about the topic of transforming an arbitrary dynamical system to a quasi-attractor system:

**Problem 13.7.** *Let $f : \Omega \subset \mathbb{R}^n \longrightarrow \mathbb{R}^n$ be a $C^r$ real function that defines a discrete map also called $f$, not necessarily of the quasi-attractor type, and $\Omega$ is a manifold. Find a controller $u$ such that the function $f + u : \Omega' \subset \mathbb{R}^n \longrightarrow \mathbb{R}^n$ defines a non-trivial quasi-attractor map, or find another method to do that. Here $\Omega'$ is the domain of definition of $f + u$.*

**Problem 13.8.** *Let $f : \Omega \subset \mathbb{R}^n \longrightarrow \mathbb{R}^n$ be a $C^r$ real function that defines a continuous-time system also called $f$, not necessarily of the quasi-attractor type, and $\Omega$ is a manifold. Find a controller $u$ such that the* *function* $f + u : \Omega' \subset \mathbb{R}^n \longrightarrow \mathbb{R}^n$ *defines a non-trivial quasi-attractor continuous-time system, or find another method to do that. Here* $\Omega'$ *is the domain of definition of* $f + u$.

**Problem 13.9.** *In particular, let* $f : \Omega \subset \mathbb{R}^n \longrightarrow \mathbb{R}^n$ *be a* $C^r$ *real function that defines a discrete quasi-attractor map also called* $f$, *and* $\Omega$ *is a manifold. Find a controller* $u$ *such that the function* $f + u : \Omega' \subset \mathbb{R}^n \longrightarrow \mathbb{R}^n$ *defines a non-trivial hyperbolic map in the sense of Definition 1, or find another method to transform a quasi-attractor chaotic map to a hyperbolic chaotic map, and vice versa. Here* $\Omega'$ *is the domain of definition of* $f + u$.

**Problem 13.10.** *In particular, let* $f : \Omega \subset \mathbb{R}^n \longrightarrow \mathbb{R}^n$ *be a* $C^r$ *real function that defines a continuous-time quasi-attractor system also called* $f$, *and* $\Omega$ *is a manifold. Find a controller* $u$ *such that the function* $f + u : \Omega' \subset \mathbb{R}^n \longrightarrow \mathbb{R}^n$ *defines a non-trivial hyperbolic continuous-time system in the sense of Definition 2, or find another method to transform a quasi-attractor chaotic continuous-time system to a hyperbolic chaotic continuous-time system, and vice versa. Here* $\Omega'$ *is the domain of definition of* $f + u$.

**Problem 13.11.** *In particular, let* $f : \Omega \subset \mathbb{R}^n \longrightarrow \mathbb{R}^n$ *be a* $C^r$ *real function that defines a discrete quasi-attractor map also called* $f$, *and* $\Omega$ *is a manifold. Find a controller* $u$ *such that the function* $f + u : \Omega' \subset \mathbb{R}^n \longrightarrow \mathbb{R}^n$ *defines a non-trivial Lorenz-type chaotic map in the sense of Definition 11, or find another method to transform a quasi-attractor chaotic map to a Lorenz-type chaotic map, and vice versa. Here* $\Omega'$ *is the domain of definition of* $f + u$.

**Problem 13.12.** *In particular, let* $f : \Omega \subset \mathbb{R}^n \longrightarrow \mathbb{R}^n$ *be a* $C^r$ *real function that defines a continuous-time quasi-attractor system also called also* $f$, *and* $\Omega$ *is a manifold. Find a controller* $u$ *such that the function* $f + u : \Omega' \subset \mathbb{R}^n \longrightarrow \mathbb{R}^n$ *defines a non-trivial chaotic continuous-time system of the Lorenz-type in the sense of Definition 11, or find another method to transform a chaotic continuous-time system of the quasi-attractor type to a chaotic continuous-time system of the Lorenz-type, and vice versa. Here* $\Omega'$ *is the domain of definition of* $f + u$.

## 13.5. A Common Classification of Strange Attractors of Dynamical Systems

Generally, at the present time, strange attractors can be classified into three principal classes [Anishchenko & Strelkova, 1997; Plykin, 2002]: hyperbolic,

Lorenz-type, and quasi-attractors, described with some of their principal properties in the above sections. For these classes, the chaos can be proved by finding a positive Lyapunov exponent, a continuous frequency spectrum, fast decaying correlation functions, etc. We notice that this classification is based on the rigorous mathematical analysis and is not accepted as significant from the experimental point of view. In fact, it was shown in [Anishchenko & Strelkova, 1998] that properties of these types of chaotic attractors are basically different. The presence of homoclinic tangencies of the trajectories is the most common property of quasi-attractors and Lorenz-type attractors, and the difference between them is that quasi-attractors contain a countable subset of stable periodic orbits. The Lorenz-type attractors can be studied using statistical methods because they admit the introduction of reasonable invariant measures, contrary to the quasi-attractors. Hence a rigorous foundation for studying characteristics of these attractors such as Lyapunov exponents can be employed. In this case, hyperbolic and Lorenz attractors are called *stochastic*. The presence of saddle-focus homoclinic loops in the quasi-attractors implies that this type is more complex than the hyperbolic and Lorenz-type attractors. Hence quasi-attractors are not suitable for some potential applications of chaos such as secure communications and signal masking. Hence we state the following problem:

**Problem 13.13.** *Determine the type of chaos (one of the three types defined above) resulting from the several techniques of chaotification well known in the literature.*

Of course, to solve such a problem, one may use the resulting expressions of maps (systems) resulting from the application of a specific chaotification method.

## References

1. Abraham, R. & Marsden, J. E. [1978] *Foundations of Mechanics*, Benjamin/Cummings Publishing. Reading Mass.
2. Afraimovich, V. S., Bykov, V. V. & Shil'nikov, L. P. [1982] "On structurally unstable attracting limit set of the type of Lorenz attractor," *Trans. Moscow. Math. Soc* **44**, 153–216.
3. Afraimovich, V. S., Bykov, V. V. & Shil'nikov, L. P. [1977] "On the appearance and structure of Lorenz attractor," *DAN SSSR* **234**, 336–339.
4. Afraimovich, V. S. & Shil'nikov, L. P. [1983] "Strange attractors and quasi-attractors," in *Nonlinear Dynamics and Turbulence* eds. by G.I. Barenblatt, G. Iooss &D. D. Joseph. Pitman. NY. (1983) 1–28.

5. Alves, J. F., Araújo, V., Pacifico, M. J. & Pinheiro, V. [2007] "On the volume of singular-hyperbolic sets," *Dynamical Systems* **22**(3), 249–267.
6. Anishchenko, V. [1990] *Complex Oscillations in Simple Systems*, Nauka. Moscow.
7. Anishchenko, V. [1995] *Dynamical Chaos Models and Experiments*, World Scientific. Singapore.
8. Anishchenko, V. S., Luchinsky, D. G., McClintoc,. P. I., Khovanov, A. & Khovanova, N. A. [2002] "Fluctuational escape from a quasi-hyperbolic attractor in the Lorenz system," *J. Exper. Theor. Phys*, **94**(4), 821–833.
9. Anosov, D. V. [1967] "Geodesic flows on closed Riemannian manifolds of negative curvature," *Proc. Steklov Math. Inst* **90**, 1–235.
10. Anosov, D. V., Klimenko. A. V. & Kolutsky, G. [2008] "On the hyperbolic automorphisms of the 2-torus and their Markov partitions," Preprint of Max-Plank Institute for Mathematics.
11. Araujo, V. & Pacifico, M. J. [2007] "Three Dimensional Flows," *XXV Brazillian Mathematical Colloquium*. IMPA. Rio de Janeiro.
12. Araujo, V. & Bessa, M. [2008] "Dominated splitting and zero volume for incompressible three-flows," *Nonlinearity* **21**(7), 1637–1653.
13. Araujo, V. & Pacifico, M. J. [2008] "What is new on Lorenz-like attractors," preprint. arXiv0804.3617.
14. Arnold, V. I. [1988] *Geometrical methods in the theory of ordinary differential equations*, Springer-Verlag. Berlin.
15. Arrowsmith, D. K. & Plaa, C. M. [1990] *An introduction to dynamics systems*, Cambridge University Press.
16. Banerjee, S. & Grebogi, C. [1999] "Border collision bifurcations in two-dimensional piecewise smooth maps," *Phys. Rev. E* **59**(4), 4052–4061.
17. Belykh, V. [1982] "Models of discrete systems of phase locking," In *Phase Locking Systems*. (L. N. Belyustina &V. V. Shakhgil'dyan. Eds.), Radio Svyaz. Moscow. 161–176 (in Russian).
18. Belykh, V. [1995] "Chaotic and strange attractors of two dimensional map," *Math. Sbornik* **186** (3), 311–320 (in Russian).
19. Belykh, V., Belykh, I. & Mosekilde, E. [2005] "Hyperbolic Plykin attractor can exist in neuron models," *Inter. J. Bifurcation & Chaos* **15**(11), 3567–3578.
20. Bunimovich, L. A. [2000] *Dynamical Systems. Ergodic Theory and Applications, Encyclopedia of Mathematical Sciences* **100**, Springer. New York.
21. Bykov, V. & Shil'nikov, L. P. [1989] "On the boundaries of the domain of existence of the Lorenz attractor," In *Methods of Qualitative Theory and Theory of Bifurcations*. Gorky State University. Gorky. 151–159 (in Russian).
22. Changming, D. [2005] "The omega limit sets of subsets in a metric space," *Czechoslovak Mathematical Journal*, **55** (1), 87–96.
23. Chen, G. & Dong. X. [1998] *From Chaos to Order*, World Scientific, Singapore.
24. Chen, G. & Lai. D. [1997a] "Anticontrol of chaos via feedback," In *Proc. of IEEE Confence on Decision and Control* San Diego. CA, 367–372.

25. Chen, G. & Lai, D. [1997b] "Making a discrete dynamical system chaotic: Feedback control of Lyapunov exponents for discrete-time dynamical system," *IEEE Trans.Circ. Syst.-I* **44**, 250–253.
26. Chen, G. & Lai, D. [1998] "Feedback anticontrol of chaos," *Int. J. Bifurcation & Chaos* **8**, 1585–1590.
27. Chen, G., Hsu, S. & Zhou, J. [1998] "Snap-back repellers as a cause of chaotic vibration of the wave equation with a van der Pol boundary condition and energy injection at the middle of the span," *J. Math. Phys* **39**(12), 6459–6489.
28. Devaney, R. L. [1989] *An Introduction to Chaotic Dynamical Systems.* Addison–Wesley. New York.
29. Ott, E. [1993] *Chaos in Dynamical Systems*, Cambridge Univ. Press. Cambridge.
30. Fujii, H., Aihara, K. & Tsuda, I. [2007(a)] "Corticopetal acetylcholine: A role in attentional state transitions and the genesis of quasi-attractors during perception," *Advances in Cognitive Neurodynamics ICCN.* Preprint.
31. Fujii, H., Aihara, K. & Tsuda, I. [2007(b)] "Corticopetal acetylcholine: Possible scenarios on the role for dynamic organization of quasi–attractors," Lecture Notes in Computer Science. Neural Information Processing 14th International Conference. ICONIP [2007] Kitakyushu. Japan. November 13–16 [2007] Revised Selected Papers. Part I. 170–178.
32. Gavrilov, N. K. & Shil'nikov, L. P. [1973] "On three dimensional dynamical system close to systems with a structuraly satble homoclinic curve," *Math. USSR. Sb* **19**, 139–156.
33. Guckenheimer, J. & Holmes, P. [1983] *Nonlinear Oscillations. Dynamical Systems and Bifurcations of Vector Fields.* New York. Springer Verlag.
34. Hayashi, S. [1992] "Diffeomorphisms in $C^1$ $(M)$ satisfy Axiom A," *Ergod. Th. and Dynam. Sys* **12**, 233–253.
35. Hénon, M. [1976] "A two dimensional mapping with a strange attractor," *Commun. Math. Phys* **50**, 69–77.
36. Hunt, T. J. [2000] "Low dimensional dynamics bifurcations of cantori and realisations of uniform hyperbolicity," PhD thesis. Univ. of Cambridge. URL: http//www.timhunt.me.uk/maths/thesis.ps.gz.
37. Isaeva, O. V., Jalnine, A. Yu. & Kuznetsov, S. P. [2006] "Arnold's cat map dynamics in a system of coupled non-autonomous van der Pol oscillators," *Phys. Rev. E* **74**, 046207.
38. Katok, A. & Hasselblatt, B. [1995] *Introduction to the Modern Theory of Dynamical Systems*, Cambridge University Press.
39. Klinshpont, N. E., Sataev, E. A. & Plykin, R. V. [2005] "Geometrical and dynamical properties of Lorenz type system," *J. Phys. Confer. Series* **23**, 96–104.
40. Bautista, S. & Morales, C. [2006] "Existence of periodic orbits for singular-hyperbolic sets," *Moscow Mathematical Journal* **6**(2), 265–297.
41. Klinshpont, N. E. [2006] "On the problem of topological classification of Lorenz-type attractors," *Math. Sbornik* **197**(4), 75–122.
42. Komuro, M. [1984] "Expansive properties of Lorenz attractors," In *The the-*

*ory of dynamical systems and its applications to nonlinear problems* 4–26. World Sci. Publishing. Kyoto.
43. Kupka, I. [1964] "Contribution à la théorie des champs génériques," *Contributions to Differential Equations* **2**, 457–484 and **3**, 411–420.
44. Kuznetsov, S. & Seleznev, E. [2006] "A strange attractor of the Smale-Williams type in the chaotic dynamics of a Physical system," *J. Exper. Theor. Physics* **102**(2), 355–364.
45. Kuznetsov, S. P. [2005] "Example of a Physical System with a Hyperbolic Attractor of the Smale-Williams Type," *Phys. Rev. Lett* **95**, 144101.
46. Kuznetsov, S. P. & Pikovsky, A. [2007] "Autonomous coupled oscillators with hyperbolic strange attractors," *Phys. D* **232**, 87–102.
47. Kuznetsov, S. P. [2008] "On the Feasibility of a parametric generator of hyperbolic chaos," *J. Exper. Theor. Physics* **106**(2), 380–387.
48. Kuznetsov, S. P. & Pikovsky, A. [2008] "Hyperbolic chaos in the phase dynamics of a $Q$-switched oscillator with delayed nonlinear feedbacks," *Eur. Phys. Lett* **84**, 10013.
49. Kuznetsov, S. P. & Ponomarenko, V. I. [2008] "Realization of a strange attractor of the Smale-Williams type in a radiotechnical delay-fedback oscillator," *Technical Physics Letters* **34**(9), 771–773.
50. Lai, D. & Chen. G. [2003] "Making a discrete dynamical system chaotic: Theorical results and numerical simulations," *Int. J. Bifuration & Chaos* **13**(11), 3437–3442.
51. Li, T. Y. & Yorke, J. A. [1975] "Period three implies chaos," *Amer. Math. Monthly* **82**, 481–485.
52. Li, X., Chen, G., Chen, Z. & Yuan, Z. [2002a] "Chaotifying linear Elman networks," *IEEE Trans. Neural Networks* **13**, 1193–1199.
53. Li, Z., Park, J. B., Joo, Y. H., Choi, Y. H. & Chen, G. [2002b] "Anticontrol of chaos for discrete TS fuzzy systems," *IEEE Trans. Circ. Syst.-I* **49**, 249–253.
54. Li, Z., Park, J. B., Chen, G. & Joo, Y. H.[2002c] "Generating chaos via feedback control from a stable TS fuzzy system through a sinusoidal nonlinearity," *Int. J. Bifuration & Chaos* **12,** 2283–2291.
55. Li, C. [2004] "On super-chaotifying discrete dynamical systems," *Chaos, Solitons & Fractals* **21**, 855–861.
56. Lin, W., Ruan, J. & Zhou, W. [2002] "On the mathematical clarification of the snapback repeller in high-dimensional systems and chaos in a discrete neural network model," *Int. J. Bifurcation & Chaos*, **12** (5), 1129–1139.
57. Lichtenberg, A. & Lieberman, M. [1983] *Regular and Stochastic Motion*, Springer-Verlag.
58. Lorenz, E. N. [1963] "Deterministic Non–periodic Flow," *J. Atmos. Sci* **20**, 130–141.
59. Lozi, R. [1978] "Un attracteur étrange (?) du type attracteur de Hénon," *J. Phys. (Paris)* **39**, Colloq. C5 9–10.
60. Luchinski, D. G. & Khovanov, I. A. [1999] "Fluctuation-induced escape from the basin of attraction of a quasiattractor," *JETP. Letters* **69**(11) 825–830.
61. Lû, J., Zhou, T., Chen, G. & Yang, X. [2002] "Generating chaos with a switching piecewise-linear controller," *Chaos* **12** (2), 344–349.

62. Manning, A. [1974] "There are no new Anosov diffeomorphisms on tori," *Amer. J. Math* **96**, 422–429.
63. Mañé, R. [1982] "An ergodic closing lemma," *Ann. of Math* **116**, 503–540.
64. Marotto, F. R. [1978] "Snap-back repellers imply chaos in $\mathbb{R}^n$," *J. Math. Anal. Appl* **3**, 199–223.
65. Marotto, F. R. [2005] "On redefining a snap-back repeller," *Chaos, Solitons & Fractals* **25**, 25–28.
66. Mather, J. N. [1968] "Characterization of Anosov diffeomorphisms," *Nederl. Akad. Wetensch. Proc. Ser. A.71* Indag. Math. **30**, 479–483.
67. Matsumoto, T., Chua, L. O. & Komuro, M. [1985] "The double scroll," *IEEE Trans. Circuits Syst* **CAS–32**, 797–818.
68. Mira, C. [1997] "Chua's circuit and the qualitative theory of dynamical systems," *Inter. J. Bifuration & Chaos* **7**(9), 1911–1916.
69. Mischaikow, K. & Mrozek, M. [1995] "Chaos in the Lorenz Equations: A Computer-Assisted Proof," *Bull. Amer. Math. Soc* **32**, 66–72.
70. Mischaikow, K. & Mrozek, M. [1998] "Chaos in the Lorenz equations: A computer assisted proof," Part II. Detail. *Math. Comp* **67**(223), 1023–1046.
71. Misiurewicz, M. [1980] "Strange attractor for the Lozi mapping," *Ann. N.Y. Acad. Sci* **357**, 348–358.
72. Morales, C. M., Pacifico, J. & Pujals, E. [1998] "On $C^1$ robust singular transitive sets for three–dimensional flows," *C. R. Acad. Sci. Paris. Série I* **326**, 81–86.
73. Neimark, Y. & Landa, P. [1989] *Stochastic and Chaotic Oscillations*, Nauka. Moscow.
74. Newhouse, S. [1980] "Asymptotic behavior and homoclinic points in nonlinear systems," *Ann. of N.Y. Acad. Sci* **357**, 292–299.
75. Plykin, R. V. [2002] "On the problem of topological classification of strange attractors of dynamical systems," *Russ. Math. Surv* **576**, 1163–1205.
76. Pujals, E. R., Robert, L. & Shub. M. [2006] "Expanding maps of the circle rerevisited, Positive Lyapunov exponents in a rich family," *Ergod. Th. Dynam. Syst* **26**, 1931–1937.
77. Pujals, E. R. [2008] "Density of hyperbolicity and homoclinic bifurcations for topologically hyperbolic sets," *Discrete and Continuous Dynamical System* **20**(2), 337–408.
78. Rabinovich, M. & Trubetskov, D. [1984] *The Introduction to the Theory of Oscillations and Waves*, Nauka. Moscow.
79. Ruelle, D. & Takens, F. [1971] "On the nature of turbulence," *Comm. Math. Phys*, **20**, 167–192.
80. Schuster, H. [1984] *Deterministic Chaos*, Physik-Verlag GmbH. Weinheim (F.R.G).
81. Shil'nikov, L. P. [1965] "A case of the existence of a countable number of periodic motions," *Sov. Math. Docklady* **6**, 163–166 (translated by S. Puckette).
82. Shil'nikov, L. P. [1981] "The bifurcation theory and quasi–hyperbiloc attractors," *Uspehi Mat. Nauk* **36**, 240–241.

83. Sinai, Y. [1979] "Stochasticity of Dynamical Systems," In *Nonlinear Waves.* edited by Gaponov-Grekhov. A. V. Nauka. Moscow, 192 (in Russian).
84. Smale, S. [1963] "Stable manifolds for differential equations and diffeomorphisms," *Ann. Scuola Norm. Sup. Pisa* **17**, 97–116.
85. Smale, S. [1967] "Differentiable dynamical systems," *Bull. Amer. Math. Soc* **73**, 747–817.
86. Wang, X. F & Chen, G. [1999] "On feedback anticontrol of discrete chaos," *Int. J. Bifurcation & Chaos* **9**, 1435–1441.
87. Wang, X. F. & Chen, G. [2000] "Chaotifying a stable map via smooth small amplitude high-frequency feedback control," *Int. J. Circ. Theory Appl* **28**, 305–312.
88. Wang, X., Chen, G. & Yu, X. [2000] "Anticontrol of chaos in continuous-time systems via time-delayed feedback," *Chaos* **10**, 771–779.
89. Wang, X. M. & Fang, Z. J. [2006] "The properties of borderlines in discontinuous conservative systems," *The European. Phys. D* **37**(2), 247–253.
90. Zuo, C. & Wang, X. [2007] "Attractors and quasi-attractors of a flow," *J. Appl. Math. Comput* **23**(1-2), 411–417.
91. Yang, R., Hong, Y., Qin, H. & Chen, G. [2005] "Anticontrol of chaos for dynamic systems in *p*-normal form: A homogeneity-based approach," *Chaos, Solitons & Fractals* **25**, 687–697.
92. Yoccoz, J. C. [1993] "Introduction to Hyperbolic Dynamics. Real and complex dynamical systems," *Proc. NATO Advanced Study Institute held in Hillerod.* June 20-July 2, 265–291. Edited by Bodil Branner & Paul Hjorth. NATO Advanced Science Institutes Series CMathematical and Physical Sciences. **464**. Kluwer Academic Publishers. Dordrecht (1995).
93. Zeraoulia, E. & Sprott, J. C. [2010] "Chaotifying 2-D linear maps via a piecewise linear controller function," *Nonlinear Oscillations.* In press.
94. Zeraoulia, E. & Sprott, J. C. [2010] *2-D quadratic maps and 3-D ODE's systems: A Rigorous Approach,* World Scientific Series on Nonlinear Science Series A, **73**, ISBN: 978-981-4307-74-1, 981-4307-74-2. (http://www.worldscibooks.com/chaos/7774.html).
95. Zeraoulia, E. [2011] *The Lozi map-The power of chaos: A Practical Guide for Studying Chaotic Systems* (400 pages). In preparation.

## Author Index

## Subject Index